INTERNET FUNDAMENTALS

By

Adesh Kumar Pandey

Deptt. of Computer Science

(Krishna Institute of Engg. and Tech.)

Ghaziabad (U.P.)

S.K. KATARIA & SONS®

Publisher of Engineering Books

4885/109, Prakash Mahal, Dr. Subhash Bhargav Lane,

Opposite Delhi Medical Association, Daryaganj, New Delhi-110002 (INDIA)

Phone: +91-11-43551243, Mobile: +91-9871775858, 7678425209

e-mail: katariabook@gmail.com, katariabooks@yahoo.com

Website: www.skkatariaandsons.com

(e-books, subscriptions and bulk buying also available on our website)

Published by:

S.K. Kataria & Sons®

4885/109, Prakash Mahal, Dr. Subhash Bhargav Lane,
Opposite Delhi Medical Association, Daryaganj, New Delhi-110002 (INDIA)
Phone: +91-11-43551243, *Mobile:* +91-9871775858, 7678425209
e-mail: katariabook@gmail.com, katariabooks@yahoo.com
Website: www.skkatariaandsons.com

First Edition: 2010, Second Edition: 2011 Reprint: 2025

Price: ₹ 395/-

Laser Typesetting by : AISA Enterprises, Delhi

Printed at : Al-Aziz Books, Sahibabad

Preface to the Second Edition

Now a days Internet is critical factor to customer-to-business interactions, intra-business interactions, personal interactions, and business-to-business interactions. This book provides the basics to students about to Internet and it's implementation. The book is written to address undergraduate students with a quick introduction to the major issues and topics in Internet fundamentals.

The book is written in very student friendly language and provides the complete basic knowledge of Internet fundamental. The book is organized and cover introduction to implementation of Internet concepts.

Suggestions are always welcome from the readers.

Adesh Kr. Pandey
AUTHOR

Acknowledgement

The book itself is an acknowledgement to the intensity, drive and technical competence of many individuals, who have contributed to it.

I am thankful to Almighty God, without whose grace nothing is possible for any one. I am very thankful to my friends and students who have inspired me to write this book.

I extend my gratitude to my wife Gunjan, my little daughter Shreya, my mother and my in-laws for their strength and support, failing which, this book wouldn't have been complete.

I convey my thanks to publisher of this book, for their constructive suggestions.

Adesh Kr. Pandey
AUTHOR

Contents

ELECTRONIC MAIL AND INTERNET

SECTION- A

- **Electronic Mail**
- **Introduction to Network and Internet**
- **Working of Internet**

SYLLABUS

Introduction, Advantages and Disadvantages, Userids, Passwords, E-mail Addresses, Message Components, Message Composition, Mailer Features, E-mail Inner Workings, E-mail Management, Mime Types, Newsgroups, Mailing Lists, Chat Rooms. Introduction to Networks and Internet, History, Working of Internet, Internet Congestion, Internet Culture, Business Culture on Internet. Collaborative Computing and The Internet. Modes of Connecting to Internet, Internet Service Providers (ISPs), Internet Address, Standard Address, Domain Name, DNS, IP, v6 Modems and Time Continuum, Communications Software; Internet Tools.

CHAPTER 1

Electronic Mail

INSIDE THIS CHAPTER

1.1. INTRODUCTION TO E-MAIL

[MDU May 2005, 2006, 2008]

You can send e-mail from your computer at any time of the day to any address around the world and your electronic letter will arrive at its destination seconds after you send it, even if the receiver lives on the other side of the world.

E-mail is one of the most popular applications on the Net. It is a cheap and effective way to keep in touch with friends and family. To send a letter in Australia costs 50 cents (excluding the cost of paper for the envelope and letter), whereas, providing you have a computer, e-mail costs virtually nothing. To use e-mail, we need software called a "mail client". Often people use "Outlook Express" or "Eudora Pro" for Windows machines or "Claris E-mailer" for Macintosh computers.

Definition *"E-mail is as simple and straight forward as letter writing. E-mail is an electronic version of sending a letter".*

Definition *"E-mail, e-mail or electronic mail is the transmission of messages (e-mails or e-mail messages) over electronic networks like the internet".*

1.1.1. Advantages of E-mail **[MDU May 2008]**

Electronic mail, or **e-mail**, is the most frequently used service on the Internet for many reasons :

1. Send a message **anytime**, **anywhere** and the recipient can read it at his or her convenience.

2. Send the same message to **multiple recipients**.

FIGURE 1.1. Use of E-Mail.

3. **Easy to use**, E-mail frees us from the tedious task of managing data of daily use. It helps us to manage our contacts, send mails quickly, maintain our mail history, store the required information, etc.
4. **Forward information** to people without retyping it.
5. **E-mail is fast**, usually taking no more than a few minutes to be received.
6. **Attach digital files** to your messages, including electronic documents, video clips, music and photos.
7. **Send messages around the world** as easily as to someone down the block.
8. Sending and receiving e-mails is typically free or having **very low cost.**

1.1.2. Disadvantages of E-Mail [MDU May 2008]

Like other good thing e-mail also has some disadvantages as follows :

1. E-mail reaches the recipient most of the time, but **delivery is not guaranteed**. Sometimes spam filtering software may block a message or direct it to the recipient's **Junk Mail folder**.
2. There is **no guarantee** of e-mail privacy. A message can be intercepted and read as it makes its way over the Internet to its final destination.
3. Unencrypted mails are unsafe. So the safest way to send sensitive or confidential information via e-mail is by using **encryption software.**

4. Unsolicited e-mails that are sent in huge numbers, may slow mail servers as well as e-mail programs.
5. With hundreds of these junk mails in our inbox, the occasional good e-mail can be hard to spot.
6. The recipient needs to scan the mails, as **viruses** are transmitted through them and have the potential to harm computer systems.
7. Checking and deleting these unwanted mails can unnecessarily consume a lot of time, and it has become necessary to block or filter the unwanted e-mails by means of ***spam filters.*** Spamming includes, sending hoax e-mails.

1.1.3. User Id and Password

User id is the unique e-mail address through which particular user can use the e-mail system on which he/she has created e-mail id. For example **adesh@kiet.edu** is user id through which I can use the e-mail system of KIET.

Definition ***A password is a secret word or string of characters that is used for authentication, to prove identity or gain access to a e-mail system. The password must be kept secret from those not allowed access.***

If a user is the registered user on e-mail system then he/she can use his e-mail by using user id and password.

1.2. E-MAIL ADDRESSES

Definition ***An e-mail address is required to send and receive e-mail. They usually consist of three parts: a user name, an "at" symbol (@), and a domain name.***

adesh@kiet.edu

1. **User name:** adesh

   Located to the **left** of the @ symbol, a user name identifies your account on the e-mail server that handles the e-mail.
2. **@**

   Means "**at**" and separates your account name and the name of the mail server name. For example, adesh "is located at" kiet.edu.
3. **Domain Name: kiet.edu**

   The domain name usually consists of two **pieces of identifying information**. The first piece is the name of the **e-mail server** and is located to the **right** of the @ symbol. For example, **kiet** is the name of the e-mail server. The **second piece**, usually a three-letter **extension (edu)**, indicates the **top-level domain**. They are separated by **periods**, called "**dots**."

1.3. E-MAIL MESSAGE COMPONENTS **[MDU May 2007]**

The software program we use to send, receive, and manage electronic messages is called an **e-mail client**. To send a message, we enter information similar to the heading of a typical interoffice memo. The message components are as follows :

1. **To** contains the e-mail address of the **recipient**. This is a mandatory entry.

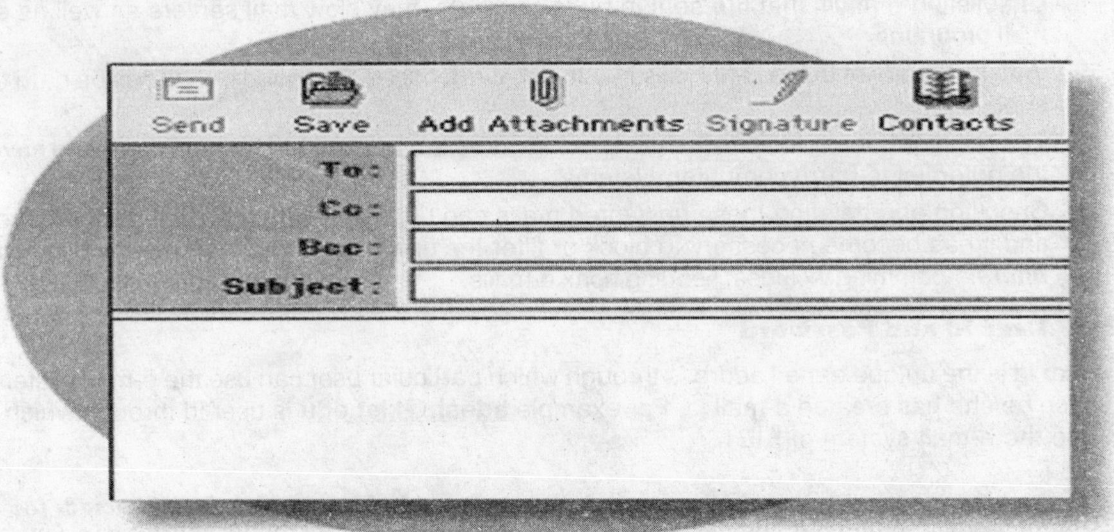

FIGURE 1.2. E-Mail Components.

2. **CC**, short for **Carbon Copy**, contains the e-mail addresses for people other than the primary recipients. This is optional.
3. **BCC**, short for **Blind Carbon Copy**, contains the e-mail addresses of other recipients who receive copies, but their names and addresses are hidden from the other recipients. This is optional.
4. **Subject** contains the main topic of the message. Recipients see this in their summary of incoming messages.
5. **Attachment** contains the names of files that you may be sending, for example, a word-processing document or a photo.
6. **Body** contains the message itself, which can be of any length.

1.4. INNER WORKING OF E-MAIL ON INTERNET

Like a postal address, an **e-mail address** specifies the **destination** of an electronic message. The **user name** is a unique name that identifies the **recipient**. The **domain name** is the address. Many people can share the same domain name. The working of e-mail on internet can be understood by following points :

- E-mail is sent and received through electronic "post offices" known as **mail server** (A computer equipped with server software) .
- To read your e-mail, you download it from the **mail server**.
- Once you enter the address of the recipient, compose your message, and click **Send**, your e-mail software handles the delivery.
- If the message isn't delivered, because of an incorrect address for example, you typically receive a message explaining why.

FIGURE 1.3. Working of E-Mail.

1.5. WRITING/COMPOSING E-MAIL

Most e-mail programs have **many features in common**. Once you know one program, you can easily use others. We can write and send the e-mail message by adopting the following steps :

1. Launching your e-mail program is usually as simple as clicking on the **e-mail icon** on your desktop. Once the program is open, write a message by clicking on the **Compose Mail**, **New Message** or similar button.

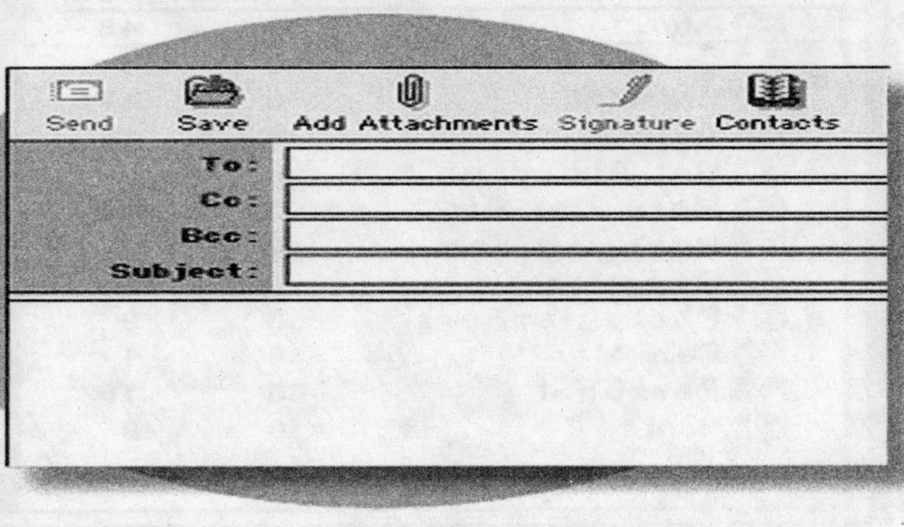

FIGURE 1.4. Writing E-Mail.

2. In the new message window, type the **e-mail addresses of the recipients** in the **To** field, or use the **address book**—a handy feature of your e-mail program that stores frequently used addresses—to select one or more e-mail addresses.
3. In the **CC** field, enter the address of anyone you want to receive a copy of the message. Type the e-mail address or choose it from your address book.
4. Type the **subject** of the message now or wait until after you've composed your message.
5. Now **type your message**. You can edit it as you would a word-processing document. You can also **copy text** from another message or document and paste it into the message box.
6. Finally, click on the **Send** button. If you have your spellchecker program enabled, it should run before the message goes off into cyberspace, allowing you to correct any errors.

1.6. READING THE E-MAIL MESSAGE

When you open your e-mail program, it lists the messages in your Inbox. From here you can read, reply to, and manage your e-mail by adopting the following steps:

1. Messages are organized in **folders** or **boxes**. Received mail is typically kept in an **Inbox**; outgoing mail is stored in an **Outbox**; sent mail is filed in a **Sent** box, and discarded mail may be kept in a **Deleted Mail** box.

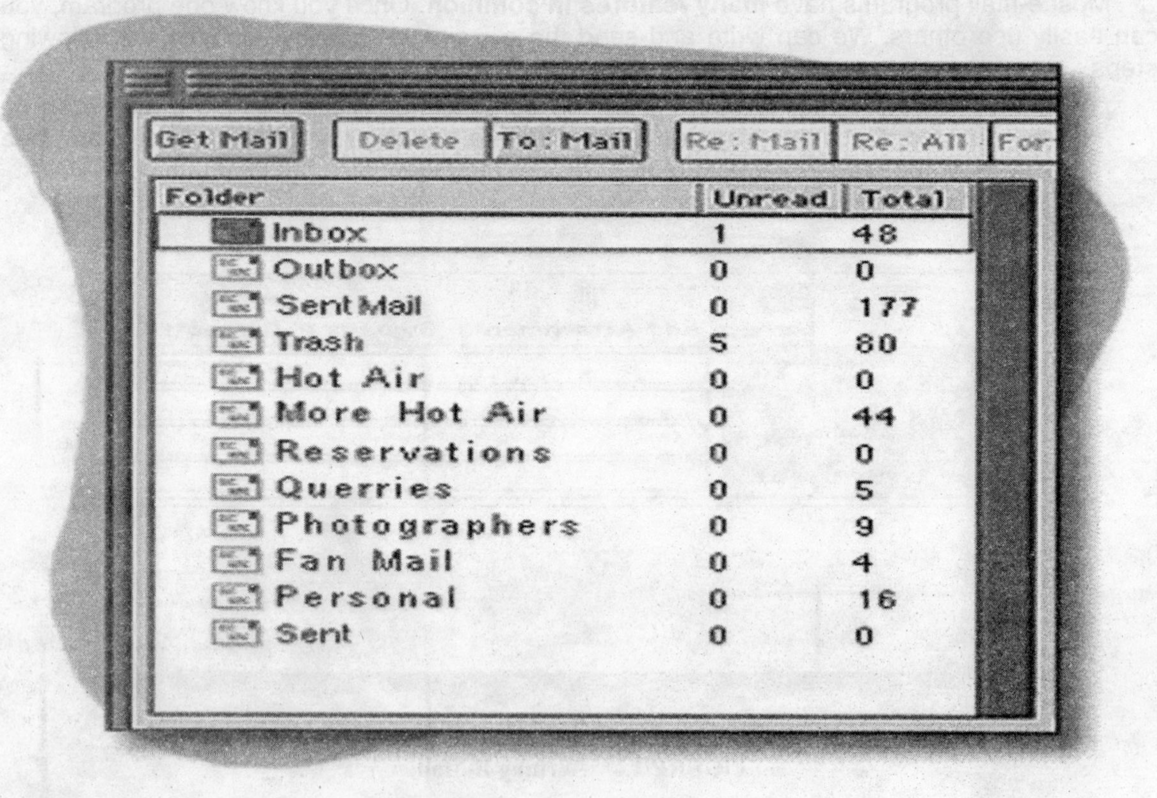

FIGURE 1.5. E-mail Message.

1. To read an e-mail, you typically double-click on the **envelope icon** or the **Subject** line of the message. This opens a new window containing the **header** and **body** of the message.
2. To review mail in other folders or boxes, click or (double-click) on the folder. Then click the message of your interest.

1.7. REPLYING TO E-MAIL MESSAGE [MDU Dec. 2006]

After we've read a message, we may want to respond. We can do as follows :

1. The window with the e-mail you are reading typically has a **Reply** button. Click this button to open a new window. The **To** and **Subject** lines are automatically filled in with the e-mail address of the sender and the subject of the message.
2. You might also have a **Reply To All** or similar button that also opens a new window. In this case, all the addresses in the **CC** list are included in the **To** and **CC** lines.
3. In some programs, the original message appears automatically. Other programs have a **Quote** or **Include Original Message** button. By clicking on this button, the text of the original e-mail appears in the body of the reply, usually **marked in a different font or color**, or with a **special character**, such as the greater than sign (>) at the beginning of each line.
4. You can choose to place your reply before or after the original text.

1.8. SENDING ATTACHMENTS

In addition to sending text messages, most e-mail programs allow you to attach word-processing documents, spreadsheets, graphics, audio, video, and other electronic files to your message.

We can attach a file as follows :

1. Click on the **Attachment** or **Include File** icon, often represented by a **paper clip**. This feature is often located on the toolbar or on a pull-down menu.
2. Locate the file on your computer's hard drive or other storage device from the window that opens requesting the name of the file.
3. Select the file you want from the file attachment window. An icon representing the file may appear in the body of your message, or the name of the file appears in the **Attachment** line of your message window.

1.9. OPENING ATTACHMENTS

To view an attached file, click on the file icon or save the file to your hard drive, then use the appropriate software application to open it.

(a) With most e-mail software, you open an attachment by **double-clicking on the icon** for the file. This typically brings up a dialog box allowing you to decide what you want to do with the file.

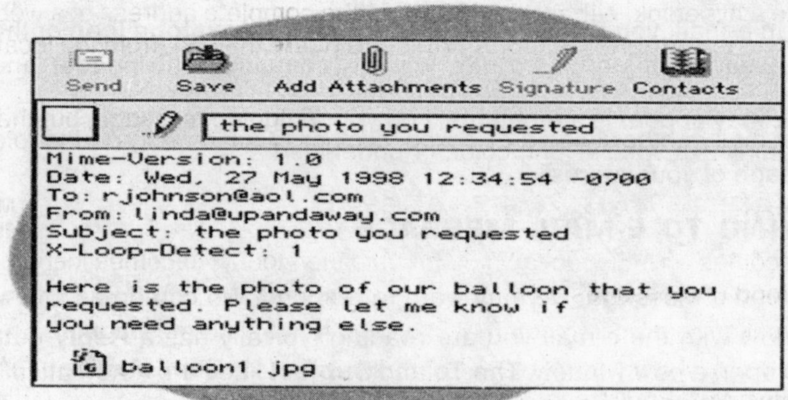

FIGURE 1.6. Opening an Attachment.

(b) Attached files must be **decoded**. This usually happens automatically.

(c) If you open an attachment that looks like **random characters and symbols**, it probably has not been decoded correctly or you may be viewing it with the wrong software application.

(d) **Use Caution :** Your computer can contract a **virus** when you open an attached file. If you do not know or trust the person who sent the e-mail, **delete the attachment** to avoid infection.

1.10. EMBEDDING A WEB LINK IN A MESSAGE

Some e-mail programs let you put a link to a **web page** in the message, making it easy for the recipient to access Web resources.

(a) Most e-mail programs support **hyperlinks** imbedding in the message text.

(b) If **hyperlinks** are supported, the recipient can just **click the link** and a web browser opens to the referenced page.

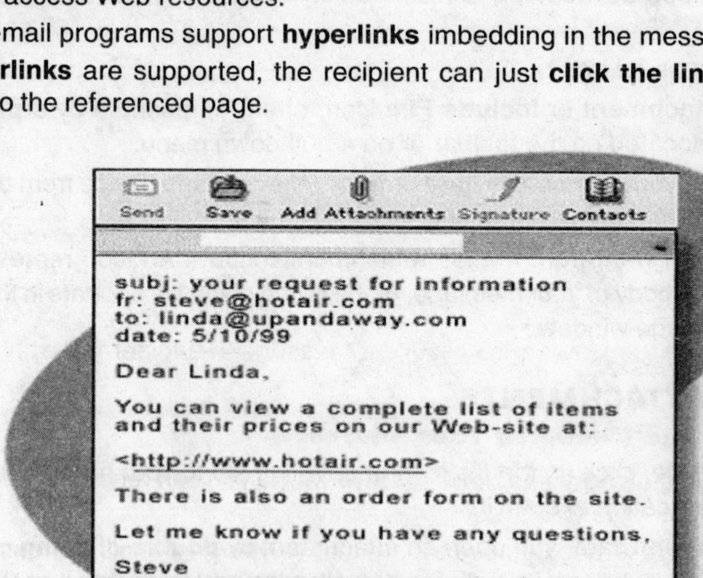

FIGURE 1.7. Embedding a Web Link.

(c) To create a hyperlink, either **type the URL** (the complete address of a web page, such as **http://www.learnthenet.com/**) or **copy and paste** the URL from the location box of your web browser.

(d) The **URL** may appear like regular text on your outgoing message, but the recipient sees the hyperlink in a special font, color, or underlined.

1.11. MANAGING YOUR MESSAGES [MDU Dec. 2006]

As e-mail becomes an indispensable communication tool, it becomes increasingly important to manage the flood of messages.

(a) Most e-mail programs let you create **folders** to file your messages.

(b) You can usually drag a message from the **Inbox** or **Sent Mail** box to another folder. Sometimes pop-up menus allow you to file messages.

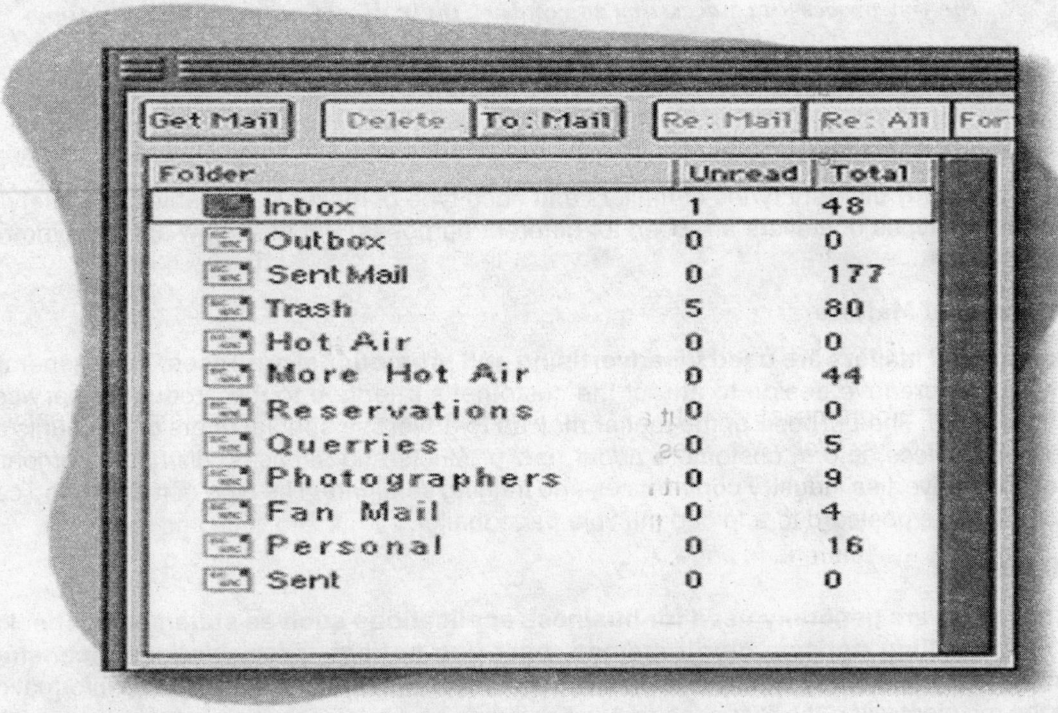

FIGURE 1.8. Managing Messages.

(c) Some e-mail programs automatically file a message in a folder you specify or discard it when it's received.

1.11.1. Setting Up and Managing Your Addresses

E-mail programs have an **address book feature** where you can record the e-mail addresses and other information of people you contact frequently.

(a) By using the address book you won't have to retype an e-mail address each time you want to send a message. This **saves time** and **avoids returned mail** due to typing mistakes.

(b) With some programs, you can store **multiple e-mail addresses** for each person and additional information, such as a company name, telephone number, and street address.

(c) With most programs, you can create a set of e-mail addresses called a **group**. When you address a message with a group, the message is automatically **addressed to all the people** in the group. This is sometimes referred to as a mailing list.

(d) As your list grows, use the **Search** or **Find** function in the address book to locate an e-mail address based on a person's first or last name.

1.12. MAILERS

Definition

A mailer is a company that sends direct mail. A mailer often employs a list broker to assist with buying mailing lists and a service bureau to assist with the list processing necessary to conduct their direct marketing campaigns.

The term mailer is sometimes used to refer to a marketer sending e-mail or other direct communications.

1.12.1. Types of Mailers

There are many different types of mailers and each type of mailer is available in a variety of styles. Different types of mailers are used for different purposes. Shown below are two types of mailers :

1. Promotional Mailers

Promotional mailers are used for advertising and promotional purposes. They generally have more of a creative design to attract the customer's attention to the product or services being advertised. The purpose of the mailer may be to advertise special offers or discounts on products or services, inform customers about new products and services, distribute company information or advertise industry conferences and training seminars. The type of mailer used can vary from a single postcard to a folded multiple page mailer.

2. Self Mailers

Self mailers are generally used for business applications such as statements, checks, order forms, billing notices, credit memos, past due notices, proxy notices, licensing, correspondence and many others. The mailer may be a one way mailer that contains information sent to the recipient without a response request or it may be a business reply piece only, such as a credit application or an order form from a catalog. The mailer could also be a two way mailer that sends information to the recipient and also contains a response mail piece requesting a response from the recipient. Self mailers save time and handling costs by eliminating the need for a mailing envelope, which improves the efficiency of the mailing process. The self mailer can contain a reply envelope which promotes a quicker response time from the customer.

1.12.2. Mailer Advantages

Let us see some of the advantages as follows :

- Eliminates the time and labor cost of separately folding inserts, and stuffing and addressing envelopes.
- Reduces material cost by not having a separate insert form, outgoing envelope and reply envelope.

- Increased efficiency generates faster turnaround time of documents.
- Less storage space is needed to store the mailers than needed for separate cartons of forms, outgoing envelopes and reply envelopes.
- Self Mailers can reduce the cost of postage because they are usually only one part forms, keeping the weight and thickness to a minimum.
- Once the mailer is imprinted and processing is completed, it is ready to mail. It does not need to be sent anywhere to be stuffed and sealed in an envelope.

1.13. E-MAIL MANAGEMENT

E-mail is now the de facto standard for business communication across organizations at this time. Recent research shows the extensive use of e-mails in business and its projected growth in the future presents major information management challenges and risks for many organizations. Just as any type of business information and record, e-mail must be included as part of, and adhered to the organizational standards addressing information and records:

- Capture
- Classification
- Storage
- Preservation
- Management
- Destruction

The vast quantities of e-mails held in inboxes, sent folders, and deleted item folders put the organization at risk and adversely impact the performance of e-mail servers in the organizations. E-mail servers were never designed to act as repositories for such great quantities of e-mails and move control of this information away from the organization. Without the management of e-mails, it is difficult for organizations to meet their legal preservation requirements in the event of litigation and government investigations, increasing the effort and cost in responding to e-discovery and disclosure.

E-mail management systems centrally capture e-mails created and received by employees. Using a classification scheme to manage this content, retention periods and access controls can be applied to manage e-mails. Metadata associated with the e-mails can be captured to allow this information to be managed and retrieved. E-mail active archiving is one of the most common applications for e-mail management. At their most basic, these solutions either copy or remove messages from the messaging application and some it at another location. Some approaches copy all messages coming in and out of the messaging application in real time, while others will physically remove the messages from the message store. In some of these solutions, the messages are not archived, but attachments are and are replaced with either a stub or an outright link within the message.

TIPS

These e-mail management systems, while attractive and effective, are merely part of an overall solution. Organizations need to have records management programs, consisting of organization-wide policies and procedures, staff and activities, in which these computer applications can be successfully utilized.

1.14. MIME [MDU May 2006, 2007, Dec. 2007]

Definition

MIME is a standard for describing different types of information. This was originally meant to be used to specify encoding for different types of information into text. This made it possible to send them as e-mail.

This is our collection of different types :

- Application
- Audio
- Image
- Message
- Multipart
- Text
- Video

Let us see some examples of common mime types in following Table 1.1.

TABLE 1.1. Different Type of MIME.

Type	Common File Extension	Purpose
text/html	.html	Web Page
image/png	.png	PNG-format image
image/jpeg	.jpeg	JPEG-format image
audio/mpeg	.mp3	MPEG Audio File
application/octet-stream	.exe	Best for downloads that should just be saved to disk

1.15. NEWS GROUPS

Newsgroups had their beginnings in an academic environment in 1979, when a couple of Duke University graduate students connected some computers together to exchange information with the UNIX community. At the same time, another graduate student at the University of North Carolina wrote the first version of the software used to distribute news.Newsgroups have been around almost since the dawn of the Internet, enabling scientists to post questions (and answers) to other scientists who were interested in the same subject.

Definition

A newsgroup is an Internet community which is dedicated to the discussion of a particular topic of interest.

Contrary to the "news" in the name, newsgroups are rarely about current events, and they cover everything from debates about the origins of humanity to tips on how to trim a cat's toenails. Newsgroups can be an immensely useful resource for information and community dialog, and people of all ages take advantage of the services offered by these communities.

A newsgroup is like an electronic public notice board on a particular subject. Anyone can post a message or read a message. Newsgroups provide a way to communicate with people who share your interests from all over the world. You can read a newsgroup any time and anyone can participate in a newsgroup... though not all Internet Service Providers (ISPs) have a news server or offer access to every newsgroup.

There are literally thousands of newsgroups covering virtually every topic imaginable - from computers, social issues, literature and science, to recreation, entertainment, hobbies and current affairs. In newsgroups you can find job postings, business and health care advice, announcements about events, referrals, political and religious discussions - even photos you can download.

Newsgroups are divided into categories. The categories help you determine what kind of group and what kind of postings you will find there.

Some common examples are :

- **alt.** means "alternative" and refers to newsgroups that are alternative (and sometime absurd), or simply that they don't fit into other categories.
- **com.** means "computers" and newsgroups proceeded with com. cover areas of interest to computer users or professionals
- **soc.** means "social" and cover topics like politics, community groups and social policy
- **talk.** Means "talk" .
- **sci.** means "science" .

We can use newsgroups for various purposes for example :

- Newsgroups are an excellent way to find out good websites to visit in your particular area of interest or just pick up detailed information about your area of interest.
- You can buy and sell stuff. People often advertise things for sale in some newsgroups.
- Gauge public opinion or strength of feeling about certain topics or interest
- Find out detailed information that the mainstream media tend to over simplify

1.16. MAILING LIST [MDU May 2005]

Definition

A list of e-mail addresses identified by a single name, such as mail-list@kiet.com. When an e-mail message is sent to the mailing list name, it is automatically forwarded to all the addresses in the list.

Most e-mail clients support mailing lists, which enables you to broadcast e-mail messages to groups that you define. In addition, there are *mailing list servers* that manage centralized mailing lists for group of users.

Here are five advantages that can be derived from keeping a mailing list:

1. A mailing list will let you to update on your clients. Instead of letting them to walk away following a sale or following to visiting your website, you can contact them again if you are able to contain their contact information. You can inform them concerning your upcoming promos. You can give them exclusive discounts to persuade them to think about your future correspondence. You can hype up an upcoming product launch to ensure fabulous traffic come launch day.
2. A considerable and receptive mailing list will give for you some great control in finding highly profitable Joint Venture (JV) affiliations with recognized names in the business. During the conceptualization stage, a JV affiliation is at all times a "what's in it for me"

affair. Potential JV partners will continuously think about a candidate who will bring the most advantages to the partnership. A considerable and responsive mailing list you can impart with them can play as a surefire ticket to the big dance.

3. A mailing list will enormously advance your sales rate. Reports say that, at the standard, it will take at least 12 links before a visitor will decide to say "yes" to your offer. Just through a mailing list will you manage to guarantee that you can contact the client that several times.
4. Maintaining a mailing list will provide you the possibility to warm up your customers for other probable transactions. By delivering regular messages with special freebies, they'd develop to trust you, hence it'll be simpler for you to persuade them to your favor when the period arises that you will have to deliver your promotional pitch.
5. Keeping a mailing list will aid you grow good relationships with your prospects. Bear in mind, it will be simpler for an individual to purchase something from someone he is acquainted with than from someone who's a virtual stranger to him.

1.17. CHAT ROOM

Chat rooms are created to hold discussions about any topic. Some of the most common topics include: sports, politics, entertainment and education. However, there are Chat Rooms in existence for practically any topic any person would want to talk about.

Chat rooms began as a text based environment for people to communicate with each other and has since expanded to include voice and video as well as three dimensional avatars which represent you in a computer based world. Although basic text based Chat Rooms are largely becoming extinct, there are still quite a few still in existence. They are quickly becoming replaced by more exciting text based Chat Rooms which feature picture avatars or various games you can play while you chat

Definition

A chat room is a virtual place on the Internet where people from all walks of life from around the world can get together in one place and textually chat with one another. The only two things a person needs to connect to a chat room are a computer and an internet connection of any speed. There are, however, Chat Rooms that require client programs, as with IRC (Internet Relay Chat).

One of the newest types of Chat Rooms allow the user to enroll in an entire virtual world. Owing their history to expensive massive multiplayer online games, these new virtual Chat Rooms (graphical multi-user environments) feature a 3D or 2D environment where people are free to make their own characters or avatars which represent them in the virtual world. Many of these virtual Chat Rooms, such as Second Life, are free of charge. Many members choose to spend real world money to purchase costumes and other virtual items. With graphical multi-user environments, you are free to develop your character however you choose and you can even create your own home away from home in the virtual world provided you are willing to buy the necessary pieces.

1.17.1. Chat Room Etiquette

Because Chat Rooms are place where so many different types of people come together to communicate with one another, certain rules need to be in place to keep the peace. Nearly all of the larger Chat Rooms are constantly moderated for any offensive activity, but quite a few smaller

Chat Rooms lack such moderation and it is therefore up to the users to be sure that no offensive activities take place.

Many Chat Rooms restrict offensive language including obscenities, racial slurs, and sexual terms that may be offensive to certain users. Any type of advertising, flooding the same word or words over and over again, and spamming are also frowned upon. Most users will be banned from the chat room for doing such activities.

If you are visiting an online chat room that is not moderated, report any of the above activities to whoever runs the site you are using. Being active in a chat community, you can greatly enhance the chatting environment for everyone in the chat room, so it is important to be proactive with policing the rules - even if you are not a chat room moderator yourself.

Significance : Chatting online with different people using chat rooms is exciting because it allows you to meet different people who share common interests. There are those who use chat rooms to find love while others use them to share the latest game chat. Whatever your reason for using chat rooms, it is important to understand the advantages and disadvantages that come along with them so that you do not fall victim to negative situations.

Types : Single and romance chat rooms are some of the most common ones on the Internet. They are mainly established by dating websites and allow people to meet each other in the cyber world. Other types of chat rooms include webcam/video chat rooms. Although this is the riskiest form of chatting, they have become popular especially on social websites because they allow seeing your chatting buddy as you chat. Other types of chat rooms include business and gaming hobby chat rooms where the users chat about common interests.

Features : Chat rooms offered by different websites range in features and security. Besides the standard "type" and "send" feature, some allow you to share pictures and share emotions that are used to portray your emotion or sudden expression. Some chat rooms also include multi-chat/ group chats that allow two or more people to isolate from the chatting population and begin their own conversation. Some chat room websites include saving options that allow you to save your conversations so you can view them again whenever you want to.

Warning : While chat rooms may be exciting cyber locations to meet new people, they can also be a dangerous ground to run into stalkers, pedophiles and more. In simple terms, be careful. You should also be careful about chat room websites that do not have rules. The "anything goes" philosophy is a dangerous one because it means anyone can do or say anything without being held accountable. If you have children, monitor what chatting websites they use and the people they chat with.

Misconceptions : Just because a chat room is labeled "Christian" or "Children's" chat room does not necessarily mean that it cannot be a danger zone. Unfortunately, many sex offenders and pedophiles roam through chat rooms looking for underage children. As shown in TV series "To Catch a Predator," there is no guarantee that any chat room website is completely safe, regardless of the rules, regulations and security tools.

1.18. E-MAIL PROTOCOLS

Definition

Protocol is about a standard method used at each end of a communication channel, in order to properly transmit information.

In order to deal with our e-mail we must use a mail client to access a mail server. The mail client and mail server can exchange information with each other using a variety of protocols like :

1. **IMAP (Internet Message Access Protocol)** is a standard protocol for accessing e-mail from our local server. IMAP is a client/server protocol in which e-mail is received and held for us by our Internet server.
2. The **POP (Post Office Protocol 3)** protocol provides a simple, standardized way for users to access mailboxes and download messages to their computers.
3. The **SMTP (Simple Mail Transfer Protocol)** protocol is used by the Mail Transfer Agent (MTA) to deliver our e-mail to the recipient's mail server. The SMTP protocol can only be used to send e-mails, not to receive them.
4. The **HTTP protocol** is not a protocol dedicated for e-mail communications, but it can be used for accessing our mailbox. Also called web based e-mail, this protocol can be used to compose or retrieve e-mails from our account. Hotmail is a good example of using HTTP as an e-mail protocol.

1.18.1. IMAP (Internet Message Access Protocol) [MDU May 2006]

Definition

IMAP stands for Internet Message Access Protocol. It is a method of accessing electronic mail or bullet in board messages that are kept on a (possibly shared) mail server.

In other words, it permits a "client" e-mail program to access remote message stores as if they were local. For example, e-mail stored on an IMAP server can be manipulated from a desktop computer at home, a workstation at the office, and a notebook computer while travelling, **without** the need to transfer messages or files back and forth between these computers.

TIPS

IMAP's ability to access messages (both new and saved) from more than one computer has become extremely important as reliance on electronic messaging and use of multiple computers increase, but this functionality cannot be taken for granted: the widely used Post Office Protocol (POP) works best when one has only a single computer, since it was designed to support "offline" message access, wherein messages are downloaded and then deleted from the mail server.

Key goals for IMAP include :

- Be fully compatible with Internet messaging standards, e.g. MIME.
- Allow message access and management from more than one computer.
- Allow access without reliance on less efficient file access protocols.
- Provide support for "online", "offline", and "disconnected" access modes
- Support for concurrent access to shared mailboxes
- Client software needs no knowledge about the server's file store format.

1.18.2. POP (Post Office Protocol) [MDU May 2006]

Definition

Protocol provides a simple, standardized way for users to access mailboxes and download messages to their computers. When using the POP protocol all your e-mail messages will be downloaded from the mail server to your local computer.

The advantage is that once your messages are downloaded you can cut the internet connection and read your e-mail at your leisure without incurring further communication costs. On the other hand you might have transferred a lot of message (including spam or viruses) in which you are not at all interested at this point.

Like it seems everything on the internet, mail retrieval is a client-server application. The Post Office Protocol defines how your e-mail client should talk to the POP server. The POP is a very simple protocol. This makes it easy to implement, has earned the Post Office Protocol widespread adoption and makes it very robust, but it also means the Post Office Protocol provides only basic functionality.

Things that can be done via the POP include :

- Retrieve mail from an ISP and delete it from the server.
- Retrieve mail from an ISP but not delete it from the server.
- Ask whether new mail has arrived but not retrieve it.
- Peek at a few lines of a message to see whether it is worth retrieving.

1.18.3. SMTP [MDU May 2006]

Definition

SMTP is the acronym for "Simple Mail Transfer Protocol", which is the standard technology used by people around the globe when sending e-mail.

Working of SMTP

SMTP is a simple ASCII protocol. After establishing the TCP connection to port 25, the sending machine, operating as the client, waits for the receiving machine, operating as the server, to talk first. The server starts by sending a line of text giving its identity and telling whether or not it is prepared to receive mail. If it is not, the client releases the connection and tries again later.

If the server is willing to accept e-mail, the client announces whom the e-mail is coming from and whom it is going to. If such a recipient exists at the destination, the server gives the client the go-ahead message. Then the client sends the message and the server acknowledges it.

1.19. E-MAIL BOMBING AND SPAMMING [MDU May 2007]

Definition

E-mail bombing is characterized by abusers repeatedly sending an e-mail message to a particular address at a specific victim site.

In many instances, the messages will be large and constructed from meaningless data in an effort to consume additional system and network resources. Multiple accounts at the target site may be abused, increasing the denial of service impact.

Definition ***E-mail spamming is a variant of bombing; it refers to sending e-mail to hundreds or thousands of users (or to lists that expand to that many users).***

E-mail spamming can be made worse if recipients reply to the e-mail, causing all the original addresses to receive the reply. It may also occur innocently, as a result of sending a message to mailing lists and not realizing that the list explodes to thousands of users, or as a result of a responder message that is setup incorrectly.

TIPS

E-mail bombing/spamming may be combined with e-mail spoofing (which alters the identity of the account sending the e-mail), making it more difficult to determine who actually sent the e-mail.

1.19.1. Prevention

Unfortunately, at this time, there is no way to prevent e-mail bombing or spamming (other than disconnecting from the Internet), and it is impossible to predict the origin of the next attack. It is trivial to obtain access to large mailing lists or information resources that contain large volumes of e-mail addresses that will provide destination e-mail addresses for the spam.

1. Develop in-house tools to help you recognize and respond to the e-mail bombing/ spamming and so minimize the impact of such activity. The tools should increase the logging capabilities as well as check for and alert you to incoming/outgoing messages that originate from the same user or same site in a very short span of time. Once you identify the activity, you can use other in-house tools to discard the messages from the offending users or sites.
2. If your site uses a small number of e-mail servers, you may want to configure your firewall to ensure that SMTP connections from outside your firewall can be made only to your central e-mail hubs and to none of your other systems. Although this will not prevent an attack, it minimizes the number of machines available to an intruder for an SMTP-based attack (whether that attack is a e-mail spam or an attempt to break into a host).
3. Consider configuring your mail handling system(s) to deliver e-mail into file systems that have per-user quotas enabled. Doing this can minimize the impact of an e-mail bombing attack by limiting the damage to only the targeted accounts and not the entire system.
4. Educate your users to call you about e-mail bombing and spamming.
5. Do not propagate the problem by forwarding (or replying to) spammed e-mail.

SUMMARY

- This chapter covers introduction of e-mail system. After going through this, students can have fair knowledge about the e-mail system, its advantages, and disadvantages.
- The chapter also covers composition of e-mail, component of e-mail message, chat room, mailing lists, mime type, e-mail protocol, e-mail bombing, spamming and e-mail management.
- Now a days e-mail management is the very big issue in corporate and business world since e-mail communication becomes an important component of business world.

- An **e-mail address** is required to send and receive e-mail. Like a postal address, an **e-mail address** specifies the **destination** of an electronic message.
- The **user name** is a unique name that identifies the **recipient** and sender of e-mail message.
- A newsgroup is like an electronic public notice board on a particular subject.
- E-mail Protocol is about a standard method used at each end of a e-mail communication channel, in order to properly transmit e-mail message.

SHORT QUESTION ANSWERS

Q. 1. What is E-mail?

Ans. E-mail, e-mail or electronic mail is the transmission of messages (e-mails or e-mail messages) over electronic networks like the internet.

Q. 2. Define User Id and Password?

Ans. User id is the unique e-mail address through which particular user can use the e-mail system on which he/she has created e-mail id. For example adesh@kiet.edu is user id through which I can use the e-mail system of KIET.

A **password** is a secret word or string of characters that is used for authentication, to prove identity or gain access to a e-mail system. The password must be kept secret from those not allowed access.

Q. 3. What is Mail Address?

Ans. An **e-mail address** is required to send and receive e-mail. They usually consist of three parts: a **user name**, an **"at" symbol (@)**, and a **domain name**.

Q. 4. Define Mailers?

Ans. A mailer is a company that sends direct mail. A mailer often employs a list broker to assist with buying mailing lists and a service bureau to assist with the list processing necessary to conduct their direct marketing campaigns. The term mailer is sometimes used to refer to a marketer sending email or other direct communications.

Q. 5. What do you understand by MIME?

Ans. MIME is a standard for describing different types of information. This was originally meant to be used to specify encoding for different types of information into text. This made it possible to send them as e-mail.

Q. 6. What is IMAP (Internet Message Access Protocol) ?

Ans. **IMAP** is a standard protocol for accessing e-mail from our local server. IMAP is a client/server protocol in which e-mail is received and held for us by our Internet server.

Q. 7. What is POP?

Ans. The **POP (Post Office Protocol 3)** protocol provides a simple, standardized way for users to access mailboxes and download messages to their computers.

Q. 8. What is SMTP?

Ans. The **SMTP (Simple Mail Transfer Protocol)** protocol is used by the Mail Transfer Agent (MTA) to deliver our e-mail to the recipient's mail server. The SMTP protocol can only be used to send e-mails, not to receive them.

Q. 9. What is HTTP?

Ans. The **HTTP protocol** is not a protocol dedicated for e-mail communications, but it can be used for accessing our mailbox. Also called web based e-mail, this protocol can be used to compose or retrieve e-mails from our account. Hotmail is a good example of using HTTP as an e-mail protocol.

Q. 10. What do you understand by newsgroup?

Ans. A newsgroup is an Internet community which is dedicated to the discussion of a particular topic of interest. Contrary to the "news" in the name, newsgroups are rarely about current events, and they cover everything from debates about the origins of humanity to tips on how to trim a cat's toenails. Newsgroups can be an immensely useful resource for information and community dialog, and people of all ages take advantage of the services offered by these communities.

REVIEW QUESTIONS

1. What is e-mail ? Write the advantages and disadvantages of e-mail system.
2. What is user-id and password and what is the significance of these two?
3. What is e-mail address?
4. Write the various component of e-mail message.
5. How e-mail system works on internet?
6. Write the short notes on following mailer features:
   (a) Writing a e-mail message
   (b) Send e-mail
   (c) Send attachment
7. What is e-mail management?
8. Write the short notes on following :
   (a) Mime (b) Mailing list
   (c) Chat room (d) News Groups
9. What is E-mail Bombing and how can we prevent it ?
10. What is spamming and how can we prevent it?
11. What do understand by e-mail protocol? write shot note on following:
   (a) HTTP (b) SMTP
   (c) POP (d) IMAP

❑❑❑

2

CHAPTER

Introduction to Network and Internet

INSIDE THIS CHAPTER

2.1. INTRODUCTION TO COMPUTER NETWORK

Computer Networks have rapidly become an integral part of business life. They allow a business to streamline the overall operation and to interact with suppliers and customers very efficiently. In many cases computer networks are seen as the solution to every problem within business.

When we communicate, we are sharing information. This sharing can be local or remote. Between individuals, local communication usually occurs face to face, while remote communication takes place over distance. The term telecommunication, which includes telephony, telegraphy, and television, means communication at a distance. Data communications are the exchange of data between two devices via some form of transmission medium such as a wire cable.

A data communications system has five components :

1. **Message :** The message is the information (data) to be communicated. Popular forms of information includes text, numbers, pictures, audio, and video.
2. **Sender :** The sender is the device that sends the data message. It can be a computer, workstation, telephone handset, video camera, and so on.
3. **Receiver :** The receiver is the device that receives the message. It can be a computer, workstation, telephone handset, television, and so on.
4. **Transmission medium :** The transmission medium is the physical path by which a message travels from sender to receiver. Some examples of transmission media include twisted-pair wire, coaxial cable, fiber-optic cable, and radio waves.
5. **Protocol :** A protocol is a set of rules that govern data communications. It represents an agreement between the communicating devices. Without a protocol, two devices may be

connected but not communicating, just as a person speaking French cannot be understood by a person who speaks only Japanese.

2.1.1. Computer Networks

Definition

"A computer network is defined to be an interconnected collection of autonomous computers".

Interconnected means capable of exchanging information, and autonomous means, that is independent or self governing. The connections allow users to exchange data, and pool and share resources, including

- Printer
- Data
- Computing Power.

In contrast to Computer Networks, where the existence of several separate machines is usually hidden from the user, Distributed Systems are computer networks with special software that makes the existence of several computers visible to the user.

2.1.2. Types of Computer Networks

The three basic types of networks include:

- Local Area Network (LAN)
- Wide Area Network (WAN)
- Metropolitan Area Network (MAN)

Local Area Network

Definition

A Local Area Network (LAN) is a network that is confined to a relatively small area. It is generally limited to a geographic area such as a lab, school, or building.

In a typical LAN configuration, one computer is designated as the file server. It stores all of the software that controls the network, as well as the software that can be shared by the computers attached to the network. Computers connected to the file server are called workstations. The workstations can be less powerful than the file server, and they may have additional software on their hard drives. On most LANs, cables are used to connect the network interface cards in each computer.

Wide Area Network

Definition

Wide Area Networks (WANs) connect larger geographic areas, such as U.P., the India, or the world. D dicated transoceanic cabling or satellite uplinks may be used to connect this type of network.

Using a WAN, schools in India can communicate with places like Tokyo in a matter of minutes, without paying enormous phone bills. A WAN is complicated. It uses multiplexers to connect local and metropolitan networks to global communications networks like the Internet.

Metropolitan Area Network

Definition ***Metropolitan Area Network (MAN), in essence a computer network that is meant to span a whole metropolitan area.***

This concept was very popular in the early 1990s when various cities envisaged establishing such networks. In practice these networks are very rare, and today only the concept survives. This is mainly due to the increasingly private providers that develop networks as opposed to cities and councils taking the development of such networks into their own hands.The above classification is purely based on the size of the area that the network covers. In addition to the area criterion other names are common.

2.1.3. Uses of Computer Networks

The main purpose of computers is to enable people to exchange data and information. This can take the form of electronic mail (e-mail) or Intra and Extranets.

On the most basic level computer networks are great to share resources, such as printers and storage space. On the most advanced level computer networks can carry video data for remote video conferencing for example.

Business Use

In the business context computer networks can be a real asset. This includes storage and retrieval of information, reducing the need for paper traffic and record keeping in paper format, as well as rationalizing the time that staff spends producing correspondence and accounts.

In a business context the following uses are typical. They are listed in ascending order of complexity :

- Share resources such as printer.
- Share data on a central server, where all data is collected.
- Communicate via e-mail.
- Exchange information via an internal network.
- Allow all staff access to the Internet.
- Use the network and the Internet to allow staff access to business data from anywhere in the world.
- Integrate the whole business operation into a networked operation, including sales activity, stock holding, quotations, ordering raw materials, control the production process, process invoices, process all the accounts, analyze business performance, quality control.

Scientific Use

Internet is standard procedure to use networks to share data and use remote computers to carry out large computations. This includes the use of supercomputers in various locations.

Research is underway to examine the way in which networked computers can be used for the parallel processing of large computations. The ultimate aim is to produce a system that would use all the resources available on a computer network (and even the Internet) as a computing resource. The use of computers connected via a network for parallel computations is a non-trivial problem. Many researchers are developing procedures for data transfer and data handling, as well as the adaptation of algorithms for given tasks to parallel processing.

Parallel computations are also carried out by large machines that contain more than one processor. There are two types :

- Processor with local memory
- Processors with shared memory

Conceptually the difference between parallel computers and computers on a network is very small. Both exchange data, both have separate memory per processor.

2.1.4. Networking Hardware

Networking hardware includes all computers, peripherals, interface cards and other equipment needed to perform data-processing and communications within the network.

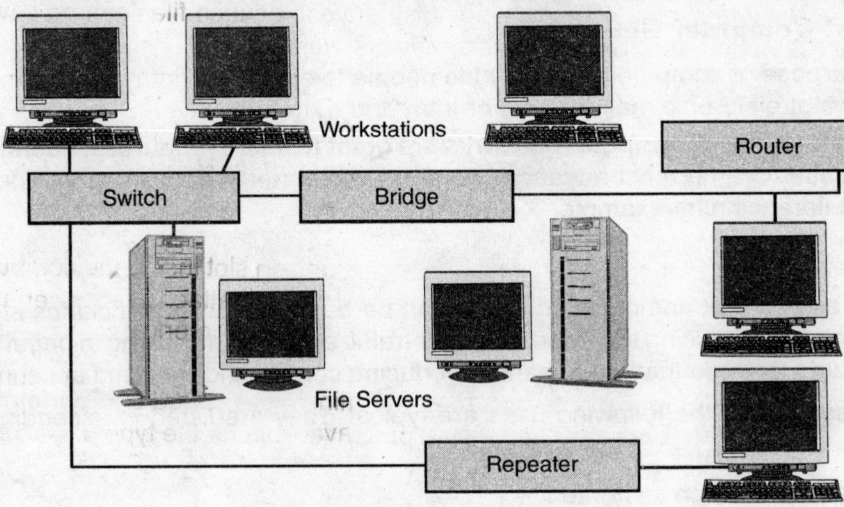

FIGURE 2.1. Network Hardware.

This section provides information on the following components:

- File Servers
- Workstations
- Network Interface Cards
- Ethernet Cards
- Token Ring Card
- Switches
- Repeaters
- Bridges
- Routers

File Servers

Definition

A file server stands at the heart of most networks. It is a very fast computer with a large amount of RAM and storage space, along with a fast network interface card. The network operating system software resides on this computer, along with any software applications and data files that need to be shared.

The file server controls the communication of information between the nodes on a network. For example, it may be asked to send a word processor program to one workstation, receive a database file from another workstation, and store an e-mail message during the same time period. This requires a computer that can store a lot of information and share it very quickly.

Workstations

Definition

All of the users connected to a network are called workstations. A typical workstation is a computer that is configured with a network interface card, networking software, and the appropriate cables.

Workstations do not necessarily need floppy disk drives because files can be saved on the file server. Almost any computer can serve as a network workstation.

Network Interface Cards

Definition

The Network Interface Card (NIC) provides the physical connection between the network and the computer workstation.

Most NICs are internal, with the card fitting into an expansion slot inside the computer. Some computers, such as Mac Classics, use external boxes which are attached to a serial port or a SCSI port. Laptop computers can now be purchased with a network interface card built-in or with network cards that slip into a PCMCIA slot.

Network interface cards are a major factor in determining the speed and performance of a network. It is a good idea to use the fastest network card available for the type of workstation you are using.

Ethernet Cards

Ethernet cards are usually purchased separately from a computer, although many computers (such as the Macintosh) now include an option for a pre-installed Ethernet card.

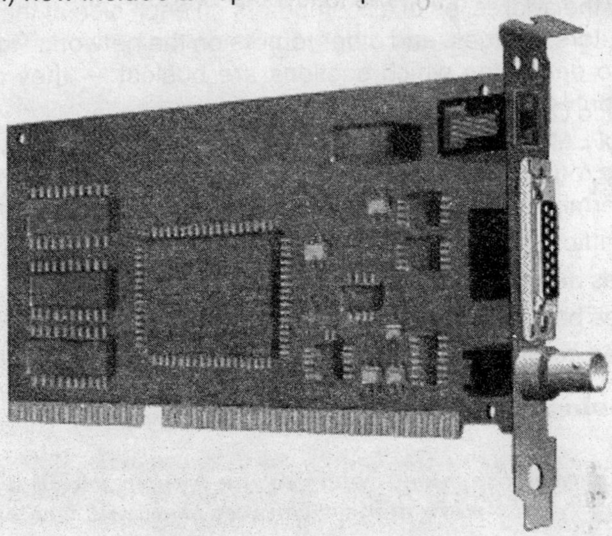

FIGURE 2.2. Ethernet Card.

Ethernet cards contain connections for either coaxial or twisted pair cables (or both) (See Fig. 2.2). If it is designed for coaxial cable, the connection will be BNC. If it is designed for twisted pair, it will have a RJ-45 connection.

Token Ring Cards

Token Ring network cards look similar to Ethernet cards. One visible difference is the type of connector on the back end of the card. Token Ring cards generally have a nine pin DIN type connector to attach the card to the network cable.

Switch

A network switch is a small hardware device that joins multiple computers together within one Local Area Network (LAN). Technically, network switches operate at layer two (Data Link Layer) of the OSI model.

Repeaters

Network repeaters regenerate incoming electrical, wireless or optical signals. With physical media like Ethernet or Wi-Fi, data transmissions can only span a limited distance before the quality of the signal degrades. Repeaters attempt to preserve signal integrity and extend the distance over which data can safely travel.

Bridges

A bridge is a device that allows you to segment a large network into two smaller, more efficient networks. If you are adding to an older wiring scheme and want the new network to be up-to-date, a bridge can connect the two.

Routers

A router translates information from one network to another; it is similar to a super intelligent bridge. Routers select the best path to route a message, based on the destination address and origin. The router can direct traffic to prevent head-on collisions, and is smart enough to know when to direct traffic along back roads and shortcuts.

While bridges know the addresses of all computers on each side of the network, routers know the addresses of computers, bridges, and other routers on the network. Routers can even "listen" to the entire network to determine which sections are busiest — they can then redirect data around those sections until they clear up.

If you have a school LAN that you want to connect to the Internet, you will need to purchase a router. In this case, the router serves as the translator between the information on your LAN and the Internet. It also determines the best route to send the data over the Internet. Routers can:

- Direct signal traffic efficiently
- Route messages between any two protocols
- Route messages between linear bus, star, and star-wired ring topologies
- Route messages across fiber optic, coaxial, and twisted-pair cabling

2.1.5. Network Topology

Definition

The term network topology refers to the way in which a network is laid out physically : two or more devices connect to a link; two or more links form a topology.

The topology of a network is the geometric representation of the relationship of all the links and linking devices (usually called nodes) to one another. There are four basic topologies possible: mesh, star, bus, and ring.

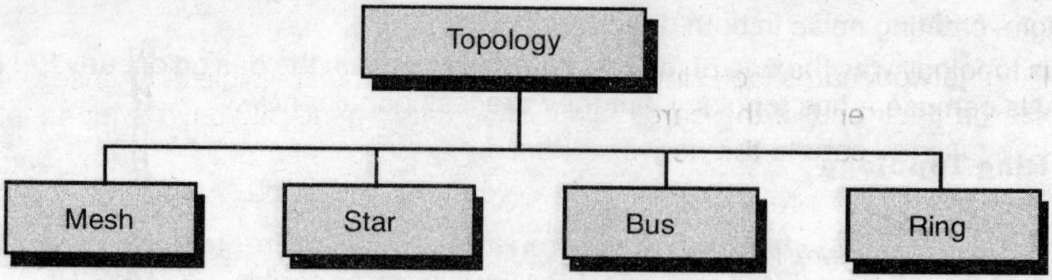

FIGURE 2.3. Categories of Topologies.

2.1.5.1. Bus Topology

Definition

A bus topology is multipoint. One long cable acts as a backbone to link all the devices in a network. Nodes are connected to the bus cable by drop lines and taps.

A drop line is a connection running between the device and the main cable. A tap is a connector that either splices into the main cable or punctures the sheathing of a cable to create a contact with the metallic core. As a signal travels along the backbone, some of its energy is transformed into heat. Therefore, it becomes weaker and weaker as it travels farther and farther. For this reason there is a limit on the number of taps a bus can support and on the distance between those taps.

FIGURE 2.4. Bus Topology.

Advantages of a Bus Topology

- Ease of installation.
- Backbone cable can be laid along the most efficient path then connected to the nodes by drop lines of various lengths. In this way, a bus uses less cabling than mesh or star topologies.

Disadvantages of a Bus Topology

- Difficult reconnection and **fault isolation.**
- A bus is usually designed to be optimally efficient at installation. It can therefore be difficult to add new devices.

- Signal reflection at the taps can cause degradation in quality.
- Adding new devices may require modification or replacement of the backbone. In addition, a fault or break in the bus cable stops all transmission, even between devices on the same side of the problem. The damaged area reflects signals back in the direction of origin, creating noise in both directions.
- Bus topology was the one of the first topologies used in the design of early LAN. Ethernet LANs can use a bus topology, but they are less popular now.

2.1.5.2. Ring Topology

Definition ***In a ring topology the devices are connected in a closed loop so that each device is connected to two others, one on either side.***

This kind of topology is robust; that is, one device's failure will probably not cause total network failure.

The most well known example of a ring topology is Token Ring.

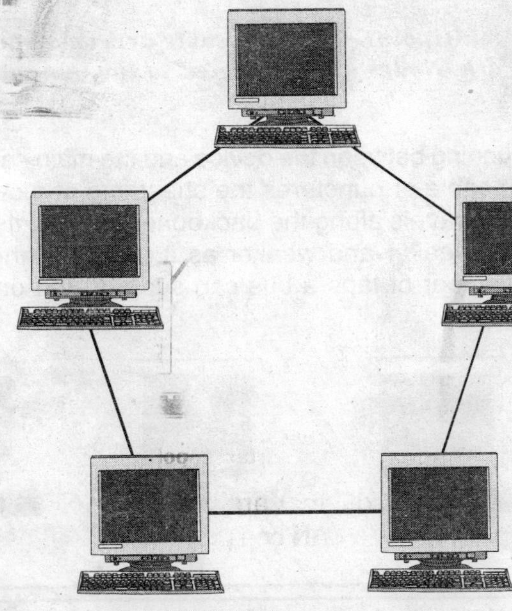

FIGURE 2.5. **Ring Topology.**

Advantages of a Ring Topology

- An orderly network where every device has access to the token and the opportunity to transmit
- Under heavy network load performs better than a start topology.
- To manage the connectivity between the computers it doesn't need network server.

Disadvantages of a Ring Topology

- One malfunctioning workstation can throw away the entire network.
- Moves, adds and changes of devices can affect the entire network.
- It is slower than an Ethernet network.

2.1.5.3. Star Topology

Definition

In a star topology the devices are all connected to a central hub, which forwards data towards its final destination.

If the data's destination is within the local star segment, the hub will forward data directly to the destination device; if the data's destination is outside the local star segment, the hub forwards the data to a router.

FIGURE 2.6. Star Topology.

Depending on the topology and media that are used, as well as the protocols (formats for transmitting data) that are implemented, a LAN can permit data transfer rates of up to 100 Million bps.

Advantages of a Star Topology

- A Star Network Topology is very easy to manage because of its simplicity in functionality.
- The problems can be easily located logically in a Star Topology and therefore is easy to troubleshoot also.
- The Star Topology is very simple in format so it is very easy to expand on the Star Topology.

Disadvantages of a Star Topology

- The Star Topology is fully dependant on the hub and the entire working of the network depends on the hub or the switch.
- If there are many nodes and the cable is long then the network may slow down.

2.1.5.4. Mesh Topology

Definition

A type of network setup where each of the computers and network devices are interconnected with one another, allowing for most transmissions to be distributed, even if one of the connections go down.

This type of topology is not commonly used for most computer networks as it is difficult and expensive to have redundant connection to every computer. However, this type of topology is commonly used for wireless networks. Below is a visual example of a simple computer setup on a network using a **mesh topology**.

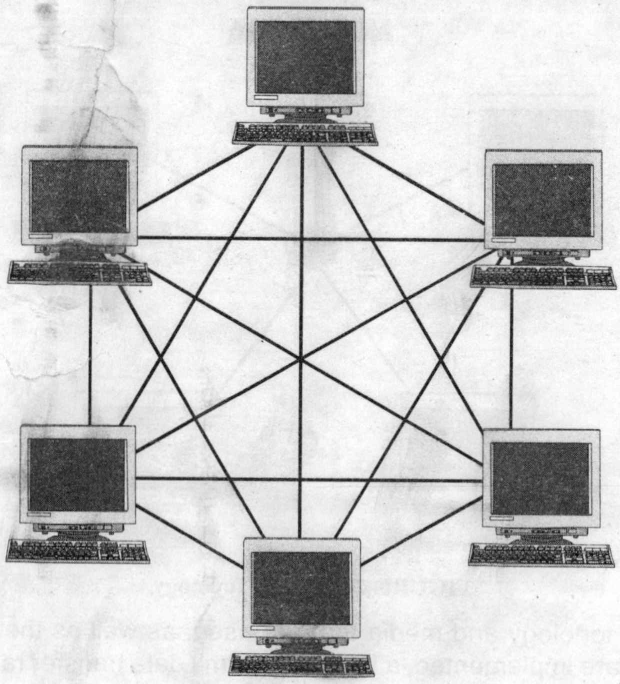

FIGURE 2.7. Mesh Topology.

The number of physical links in a fully connected mesh network with n nodes, is n(n – 1). However, if each physical link allows communication in both directions (duplex mode), we can divide the number of links by 2. In other words, we can say that in a mesh topology, we need n(n – 1)/2 links.

Advantages of a Mesh Topology

- A mesh offers several advantages over other network topologies. First, the use of dedicated links guarantees that each connection can carry its own data load, thus eliminating the traffic problems.
- Second, a mesh topology is robust. If one link becomes unusable, it does not incapacitate the entire system.
- Third, there is the advantage of privacy or security.

Disadvantages of a Mesh Topology

- The main disadvantages of a mesh are related to the amount of cabling and the number of I/O ports required.
- Every device must be connected to every other device, installation and reconnection are difficult.
- The sheer bulk of the wiring can be greater than the available space (in walls, ceilings, or floors) can accommodate.
- Finally, the hardware required to connect each link (I/O ports and cable) can be expensive. For these reasons a mesh topology is usually implemented in a limited fashion.

2.2. DEFINITION OF INTERNET [MDU May 2005, 2006]

Definition

The Internet is a network of networks, linking computers to computers sharing the TCP/IP protocols. Each runs software to provide or "serve" information and/or to access and view information.

Internet is not the only network of networks but it has three components:

1. ***Network of networks***
2. ***Network with information and resources***
3. ***Users on the network***

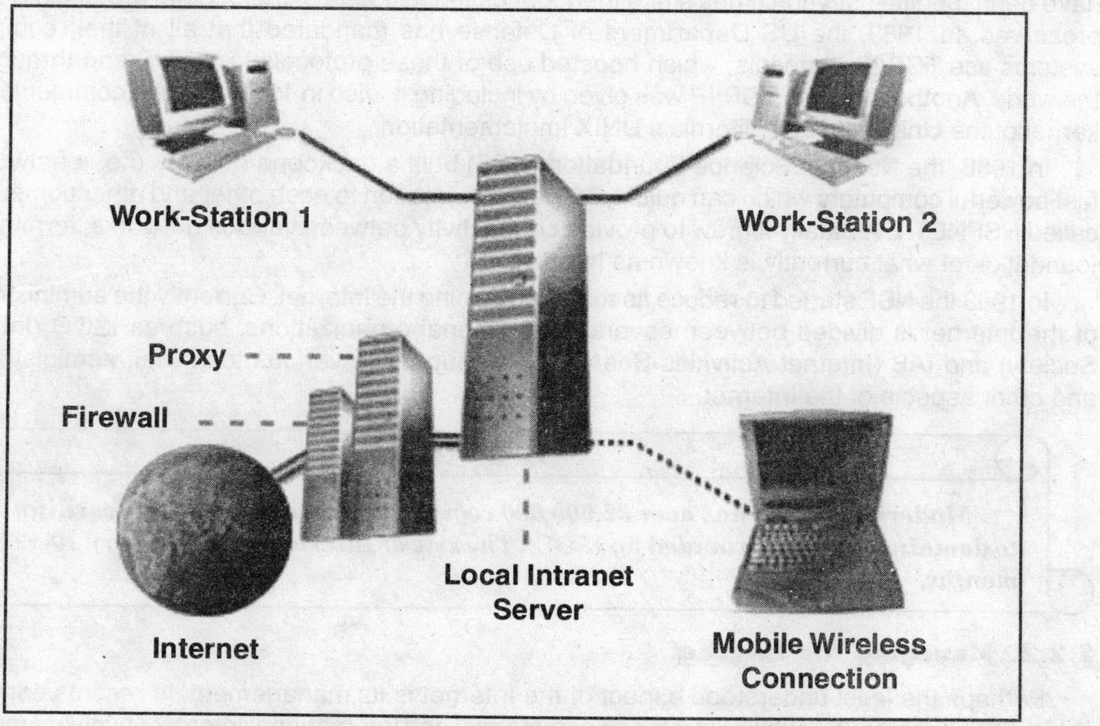

FIGURE 2.8. Internet.

Let us see some facts about Internet:

- The Internet is the transport vehicle for the information stored in files or documents on another computer.
- It can be compared to an international communications utility servicing computers. It is sometimes compared to a giant international plumbing system.
- The Internet itself does not contain information. It is a slight misstatement to say a "document was found on the Internet." It would be more correct to say it was found *through* or *using* the Internet. What it was found in (or on) is one of the computers linked to the Internet.

2.2.1. The History of Internet [MDU Dec. 2007]

Until the 1960s, very little communication was taking place between computers, and was carried out by the technology of telephone networks, i.e. circuit switching. In 1962 Paul Baran and Donald Davies independently proposed an idea of a robust, efficient, store-and-forward data network based on packets, i.e. units of data carried independently from one another. The technology was called **packet switching**, and has been implemented as ARPANET (a network of research sites in US, a project funded by the US Department of Defense) in late 1960s. In early 1970s, ARPANET spanned the continental US, and by 1973 it had connections to Europe.

ARPANET had several protocols for communications of computers one to another or with the network. In 1974, a new, more robust suite of protocols has been developed and implemented on the ARPANET based on **Transmission Control Protocol** (TCP) and **Internet Protocol** (IP). The standard includes a large collection of protocols, some of which we mention later. These protocols have been modified several times since then, but essential ideas of the original protocols are still preserved. In 1983, the US Department of Defense has mandated that all of their computer systems use TCP/IP protocols, which boosted use of these protocols in the US and throughout the world. Another boost to TCP/IP was given by including it, also in 1983, into the communication kernel of the University of California's UNIX implementation.

In 1986, the National Science Foundation (NSF) built a backbone network (i.e. a network of fast powerful computers which can quickly forward information to each other and other computers) called NSFNET. Eventually it grew to provide connectivity between various networks, forming the foundation of what currently is known as Internet.

In 1993 the NSF started to reduce its role in governing the Internet. Currently the administration of the Internet is divided between several international organizations, such as ISOC (Internet Society) and IAB (Internet Activities Board), which supervise various technical, administrative, and other aspects of the Internet.

TIPS

Modern Internet has over 32,000,000 registered domain names (according to domainstats.com/provided by ISOC). The size of Internet doubles every 10-12 months.

2.2.2. Managing the Internet

Perhaps the least understood aspect of the Internet is its management. In recent years, this subject has become the subject of intense commercial and international interest, involving multiple governments and commercial organizations, and recently congressional hearings. As issue is how the Internet will be managed in the future, and, in the process, what oversight mechanisms will insure that the public interest is adequately served.

In the 1970s, managing the Internet was easy. Since few people knew about the Internet, decisions about almost everything of real policy concern were made in the offices of DARPA. It became clear in the late 1970s, however, that more community involvement in the decision-making processes was essential. In 1979, DARPA formed the Internet Configuration Control Board (ICCB) to insure that knowledgeable members of the technical community discussed critical issues, educated people outside of DARPA about the issues, and helped others to implement the TCP/IP protocols and gateway functions. At the time, there were no companies that offered turnkey solutions to getting on the Internet. It would be another five years or so before companies like Cisco Systems were formed, and while there were no PCs yet, the only workstations available were specially built and their software was not generally configured for use with external networks; they were certainly considered expensive at the time.

In 1983, the small group of roughly twelve ICCB members was reconstituted (with some substitutions) as the Internet Activities Board (IAB), and about ten "Task Forces" were established under it to address issues in specific technical areas. The attendees at Internet Working Group meetings were invited to become members of as many of the task forces as they wished.

The management of the Domain Name System offers a kind of microcosm of issues now frequently associated with overall management of the Internet's operation and evolution. Someone had to take responsibility for overseeing the system's general operation. In particular, top-level domain names had to be selected, along with persons or organizations to manage each of them. Rules for the allocation of Internet addresses had to be established. DARPA had previously asked the late Jon Postel of the USC Information Sciences Institute to take on numerous functions related to administration of names, addresses and protocol related matters. With time, Postel assumed further responsibilities in this general area on his own, and DARPA, which was supporting the effort, gave its tacit approval. This activity was generally referred to as the Internet Assigned Numbers Authority (IANA). In time, Postel became the arbitrator of all controversial matters concerning names and addresses until his untimely death in October 1998.

TIPS

The rapid growth of the Internet has triggered the design of a new and larger address space (the so-called IP version 6 address space); today's Internet uses IP version 4. However, little momentum has yet developed to deploy IPv6 widely. Despite concerns to the contrary, the IPv4 address space will not be depleted for some time. Further, the use of Dynamic Host Configuration Protocol (DHCP) to dynamically assign IP addresses has also cut down on demand for dedicated IP addresses. Nevertheless, there is growing recognition in the Internet technical community that expansion of the address space is needed, as is the development of transition schemes that allow interoperation between IPv4 and IPv6 while migrating to IPv6.

It is helpful to consider separately the problem of managing the domain name space and the Internet address space. These two vital elements of the Internet architecture have rather different characteristics that color the management problems they generate. Domain names have semantics that numbers may not imply; and thus a means of determining who can use what names is needed. As a result, speculators on Internet names often claim large numbers of them without intent to use them other than to resell them later. Alternate resolution mechanisms, if widely adopted, could significantly change the landscape here.

In 1998, the Internet Corporation for Assigned Names and Numbers (ICANN) was formed as a private sector, non-profit, organization to oversee the orderly progression in use of Internet names and numbers, as well as certain protocol related matters that required oversight. The birth of this organization, which was selected by the Department of Commerce for this function, has been difficult, embodying as it does many of the inherent conflicts in resolving discrepancies in this arena. However, there is a clear need for an oversight mechanism for Internet domain names and numbers, separate from their day-to-day management.

Many questions about Internet management remain. They may also prove difficult to resolve quickly. Of specific concern is what role the U.S. government and indeed governments around the world need to play in its continuing operation and evolution. This is clearly a subject for another time.

2.2.3. Applications and Uses of Internet

The development of the internet is not an end in itself; it is a means by which the people can achieve a broad range of economic and social goals.

This internet can be used by all, not just by scientists and engineers. As entrepreneurs, factory workers, doctors, teachers, federal employees, and citizens, people can use this technology to :

- Create jobs, and lead towards growth.
- Reduce health care costs while increasing the quality of service in underserved areas;
- Deliver higher-quality, lower-cost government services;
- Prepare our children for the fast-paced workplace of the 21st century; and
- Build a more open and participatory democracy at all levels of government.

Let us see some of the **applications** of the Internet:

1. Economic Application

The Internet will help create high-wage jobs, stimulate economic growth, enable new products and services, and strengthen overall economy of the country. Below are some of the potential benefits to the economy:

(i) **Economic growth :** Internet can help to increase the economic growth and productivity of any country.

(ii) **Job creation :** Industry experts believe that the Personal Communications Services industry, a new family of wireless services, could create as many as 300,000 jobs in the next 10-15 years.

(iii) **Regional, state, and local economic development :** In today's knowledge-based, global economy in which capital and technology are increasingly mobile, the quality of America's information infrastructure will help determine whether companies invest here or overseas. States and regions increasingly recognize that development of their information infrastructure is key to creating jobs and attracting new businesses:

(iv) **Electronic commerce :** Electronic commerce (e.g., on-line parts catalogues, multi- media mail, electronic payment, brokering services, collaborative engineering) can dramatically reduce the time required to design, manufacture, and market new products. "Time to market" is a critical success factor in today's global marketplace.

2. Application in Health Care

The Internet can help solve global health care crisis. These problems will not be solved without comprehensive health care reform. Better use of information technology and the development of health care applications for the internet, however, can make an important contribution to reform. Experts estimate that telecommunications applications could reduce health care costs by $36 to $100 billion each year while improving quality and increasing access.

Below are some of the existing and potential applications:

- **Telemedicine :** By using telemedicine, doctors and other care givers can consult with specialists thousands of miles away; continually upgrade their education and skills; and share medical records and x-rays.
- **Personal Health Information Systems :** Every country can use computers and networks to promote self care and prevention by making health care information available 24 hours a day in a form that aids decision-making. Most people do not have the tools necessary to become an active and informed participant in their own health care.
- **Computer-Based Patient Records :** The Institute of Medicine has concluded that Computer-Based Patient Records are critical to improving the quality and reducing the cost of health care.

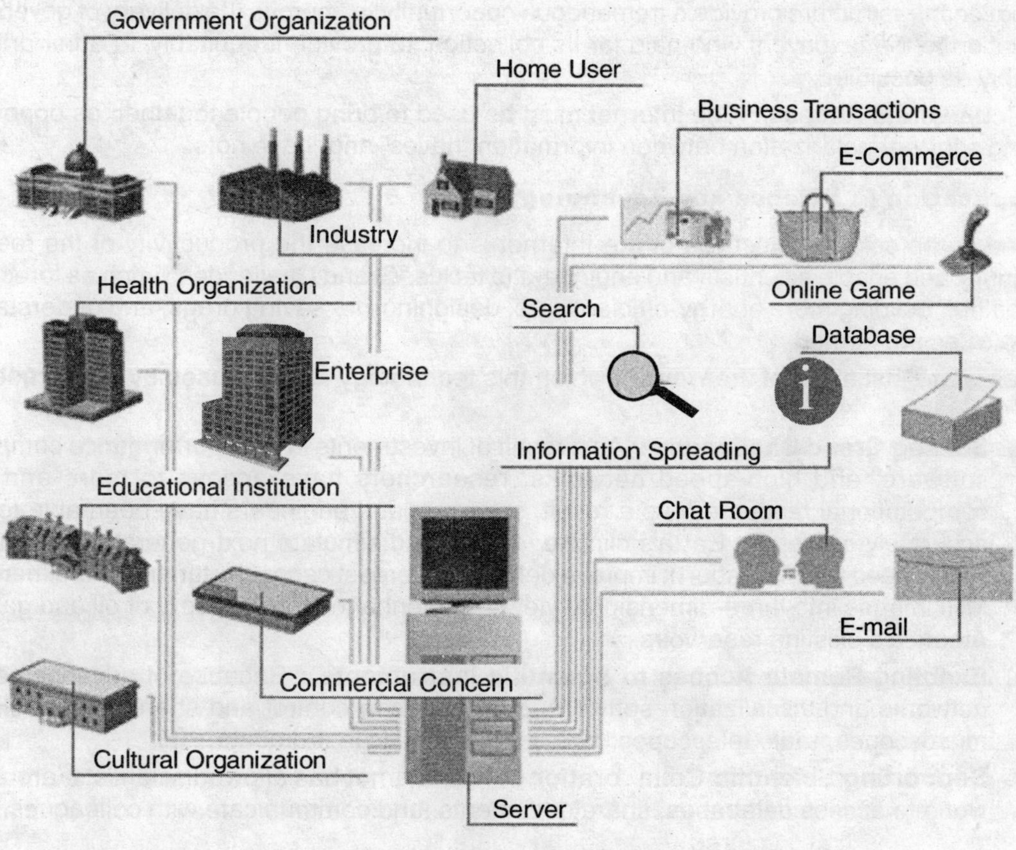

FIGURE 2.9. **Use of Internet.**

3. Application of Internet in Civic Networking

Technology in the Public Interest. The benefits of the Internet extend far beyond economic growth. As the Center for Civic Networking observed.

TIPS

"A country that works smarter; enjoys efficient, less costly government, guided by a well-informed citizenry; that produces high quality jobs and educated citizens to fill them; that paves a road away from poverty; that promotes life-long learning, public life and the cultural life of our communities. This is the promise of the Internet."

The Internet could be used to create an "electronic commons" and promote the public interest in the following ways :

(a) **Community Access Networks :** Grass-roots networks are springing up all over the country, providing citizens with a wide range of information services.

(b) **Dissemination of Government Information :** The free flow of information between the government and the public is essential to a democratic society. Improvements in the National Information Infrastructure provide a tremendous opportunity to improve the delivery of government information to the taxpayers who paid for its collection; to provide it equitably, at a fair price, as equitably as possible.

(c) **Universal Access :** The internet must be used to bring people together, as opposed to allowing a further polarization between information "haves" and "have nots."

4. Application in Science and Technology

One of the central objectives of the Internet is to increase the productivity of the research community and enable scientists and engineers to tackle "Grand Challenges," such as forecasting the weather, building more energy-efficient cars, designing life- saving drugs, and understanding how galaxies are formed.

Below are just a few of the ways in which this technology is being used by world researchers :

(a) **Solving Grand Challenges :** As a result of investments in high performance computers, software, and high-speed networks, researchers have access to more and more computational resources. As a result, scientists and engineers have been able to more accurately model the Earth's climate; design and simulate next-generation aircraft (the High Speed Civil Transport); improve detection of breast cancer by turning two-dimensional MRI images into three-dimensional views; and enhance the recovery of oil and gas from America's existing reservoirs.

(b) **Enabling Remote Access to Scientific Instruments :** Because of advancements in networks and visualization software, scientists can control and share remote electron microscopes, radio telescopes, and other scientific instruments.

(c) **Supporting Scientific Collaboration :** The Internet has allowed scientists around the world to access databases, share documents, and communicate with colleagues.

5. Application in Education

Increasingly, what we earn depends on what we learn. World must be well-educated and well-trained if want to compete internationally and enjoy a healthy democracy.

TIPS

"Imagine a biology student entering an impressive virtual laboratory environment that includes simulated molecules. The learner can pick up two molecules and attempt to fit them together, exploring docking sites. In addition to the three-dimensional images in the head-mounted display, the gesture gloves on his hands press back to provide feedback to his sense of touch. Alternatively, the student can expand a molecule to the size of a large building and fly around in it, examining the internal structure."

6. Application in E-Governance

Internet can help to provide a government, which is effective, efficient and responsive. Moving from red tape to results will require sweeping changes: emphasizing accountability for achieving results as opposed to following rules; putting customers first; empowering employees; and reengineering how government agencies do their work.

(a) **Develop a nationwide system to deliver government benefits electronically :** The government can cut costs through "electronic benefits transfer" for programs such as federal retirement, social security, unemployment insurance.

(b) **Develop integrated electronic access to government information and services :** Currently, citizen access to federal government information is uncoordinated and not customer-friendly. Electronic kiosks and computer bullet-in boards can result in quick response, complete information, and an end to telephone tag.

(c) **Establish a National Law Enforcement/Public Safety Network :** Whether responding to natural or technological disasters, or performing search and rescue or interdiction activities, federal, state, and local law enforcement and public safety workers must be able to communicate with each other effectively, efficiently, and securely.

2.2.4. Internet Issues

Emerging technologies and especially this communications revolution we are witnessing also bring with them new issues relevant to safety, privacy, security, decency and netiquette. The issues are as follows :

1. Safety

I cannot stress this section enough. Please visit these links and educate yourself and you children before you allow them to go online. Be wary of giving out your name, address or telephone number and tell them never to release this information to strangers! Above all, supervise your children while they surf! Don't assume public access Internet stations at public libraries are child-safe. Some are. Many are not.

2. Privacy

Don't post or write anything you wouldn't mind anyone else seeing. Remember your information is not secure nor is your e-mail necessarily private. It's supposed to be, but that isn't always the case. Also, never EVER send your credit card number via email.

3. Security

To enhance security, many networks have encrypted, secure transmission methods that include site certificates. Be sure you are using one of these methods if you are entering secure information, such as a credit card number at a mail-order catalog's website. Better yet, call them with this information if you are at all wary of security. There are also virtual banks, online escrow and checking systems and other Web developments emerging to deal with this issue.

4. Decency, Filters, Censorship

Unfortunately this revolution in accessing information means that not all information on the Internet is suitable for everyone. While many servers maintain standards of decency and refuse to allow any indecent material on their computer network, still many others do not, or worse, some exist solely for this purpose of allowing such material. Others oppose the material but not the principle of free speech. Look for this to be a hotly contested issue for some time. More and more, though, the loudest voices are those seeking protection for the innocent against Internet crime, including protection of children from access to pornography.

Computerized search engines that enhance our ability to find information also unfortunately make it all the more easy for children and adolescents to find information you may not want them to find. There are software programs available designed to filter out unwanted or indecent information from your computer. These programs include **Surf-Watch**, **Net Nanny** and **Cyber Patrol**.

9. Netiquette

All cultures and societies have standards of conduct and customs. There are many good resources for you to read online. The important thing is to take some time and read up on network etiquette so you can be a responsible, well-informed member of cyber-society.

2.2.5. Internet Congestion

Definition

Network congestion is the situation in which an increase in data transmissions results in a proportionately reduction, in throughput.

Throughput is the amount of data that passes through the network per unit of time, such as the number of *packets* per second. Packets are the fundamental unit of data transmission on the Internet and all other TCP/IP (Transmission Control Protocol/Internet Protocol) networks, including most LANs (Local Area Networks).

TIPS

Congestive collapse is the situation in which the congestion becomes so great that throughput drops to a low level and thus little useful communication occurs. It can be a stable state with the same intrinsic load level that would by itself not produce congestion. This is because it is caused by the aggressive retransmission used by various network protocols to compensate for the the packet loss that occurs as a result of congestion, a retransmission that continues even after the load is reduced to a level that would not have induced congestion by itself.

Congestion results from applications sending more data than the network devices (e.g., routers and switches) can accommodate, thus causing the *buffers* on such devices to fill up and possibly overflow. A buffer is a portion of a device's memory that is set aside as a temporary holding place for data that is being sent to or received from another device. This can result in delayed or lost packets, thus causing applications to retransmit the data, thereby adding more traffic and further increasing the congestion.

Network congestion is somewhat analogous to road congestion. One technique that has been used with some success to deal with road congestion is *metering*, in which the rate of vehicles entering a road or area is restricted by signals. Many economists advocate pricing (i.e., charging to use the roads, with higher prices when congested) as a more efficient congestion reduction technique. Such *congestion pricing* has become much easier to put into practice in recent years as a result of advances in electronics technology.

Various techniques have likewise been developed in attempt to minimize congestion collapse in communications networks. In addition to increasing capacity and data compression, they include protocols for informing transmitting devices about the current levels of network congestion and having them reroute or delay their transmissions according to congestion levels.

Congestion control The processes that are used to reduce congestion in a network. This includes making decisions such as: deciding when to accept new TRAFFIC, when to delete packets and when to adjust the ROUTING policies used in a network.

2.2.6. Business Culture on Internet [MDU May 2007]

Today's business environment is influenced greatly by market, economical, societal and technological factors creating a lot of competition. Also these factors are unpredictable in that they may change unexpectedly at any time. Such changes in the business world have been defined as business pressure. In order to succeed in this dynamic world companies must take action such as lowering costs and improving quality etc. At the same time they need to be innovative in customizing their products, creating new products and providing good customer service.

The various environmental business pressures on companies today can be grouped into three categories: Market, Societal and Technological. Strength of Internet Business Over traditional business can be more clear by the compression of both. Online purchasing and selling through e-business (wholesale and retail) Offers opportunities and advantages to the companies of all sizes engaged in different portfolios. Some of these advantages are enumerated below.

1. Sampling of products such as books, recorded lecturers, and music cassettes is possible on Web for business promotion.
2. Business houses selling their goods through catalogues, can reach additional global customers at lower cost.
3. For items subjected to frequent changes e.g., shares and bonds etc., the changes in rates/quotations can be known instantly.
4. Customer's knowledge can be enhanced on topics such as news about local event, market research, industrial report, software etc., which can be distributed easily over the Internet.
5. A closer relationship can be developed amongst business sellers whose customer base is in on the Internet.
6. For wholesalers, distributors, retailers etc., it is convenient and efficient to buy from Web sites.
7. High quality, specialized products can be easily sold on Internet by the retailers.
8. Organizations can conduct interviews through Internet to select suitable candidate(s) for their requirements.
9. Group discussions can be conducted for corporate offices, industrial houses, business firms, universities, etc.

2.2.6.1. Disadvantage of Internet Business

1. **Security :** Security issues are the primary concern and the biggest disadvantage of using the Internet for business, particularly if your business involves financial transactions.
2. **Staying connected :** Connectivity issues can also become a disadvantage, if you are using a computer that is not reliable and is prone to locking up, or "freezing."
3. **Availability :** Not everyone has Internet access, and many of your potential customers who are not able to use the Internet may actually be lost to you if you do not provide an alternate means of doing business with them.
4. **Access :** If you are going to use the Internet for business, you must consistently be able to get access, whether through a wireless, DSL or cable connection. Finding access can become a disadvantage if you travel for your business.
5. **Misunderstandings :** Written text, especially in emails, can easily be misread or misinterpreted when there is no face-to-face contact.

2.2.7. Collaborative Computing and Internet

Definition

Collaborative computing is a term describing a variety of activities where people interact with one another using desktops, laptops, palmtops, and sophisticated digital cellular phones. As computers are best at handling data and representing information, person-to-person communication is enriched by an ability to share, modify, or collaboratively create data and information.

We believe collaborative computing is the future of the Internet. The Internet will evolve from its current role as a channel for information dissemination to a person-to-person communication medium. In collaborative computing, we build, deploy, and support tools for personal interaction over the Internet.

Collaborative computing allows users to work together on documents and projects, usually in real time, by taking advantage of underlying network communication systems. Whole new categories of software have been developed for collaborative computing, and many existing applications now include features that let people work together over networks. Here are some examples:

- Application suites such as Microsoft Office and Exchange, Lotus Notes, and Novell Groupwise that provide messaging, scheduling, document coauthoring, rules-based message management, workflow routing, and discussion groups.
- Videoconferencing applications that allow users to collaborate over local networks, private WANs (wide area networks), or over the Internet. See "Videoconferencing" for more information.
- Internet collaboration tools that provide virtual meetings, group discussions, chat rooms, whiteboards, document exchange, workflow routing, and many other features. Multicasting is an enabling technology for groupware and collaborative work on the Internet that reduces bandwidth requirements. A single packet can be addressed to a group rather than having to send a packet to each member of the group. See "Multicasting" for more details.
- Instant messaging is like electronic mail that happens in real time. When you log onto a company server or an Internet site that tracks your "presence," other people whom you designate are notified that you are online and available for instant messaging. They can send you a message that appears immediately on your screen. You can send an immediate reply. Instant message has huge potential in many applications. Companies can use it to

form ad hoc meetings or organize meetings. It is even being used as a signaling protocol to indicate when a person changes their physical location so that Internet phone calls can be redirected to that location. Refer to "Instant Messaging" for more details.

- Workflow management is about coordinating the flow of documents (invoices, reports, legal documents, etc.) within an organization from one person to another. Refer to "Workflow Management" for more details.

TIPS

A good example of collaborative applications designed for Internet use are Microsoft's NetShow and NetMeeting. NetShow is a multimedia Web technology that enhances broadcasting with interactive content like audio, illustrated audio (images and sound), and video. This is delivered using multicast technology, which means that multiple users can receive the same broadcast at the same time. NetMeeting is a multimedia conferencing tool that works over the Internet or intranets. It provides what are essentially videophone services between users.

2.3. INTRANET

Definition

An Intranet is a network based on the internet TCP/IP open standard. An intranet belongs to an organization, and is designed to be accessible only by the organization's members, employees, or others with authorization.

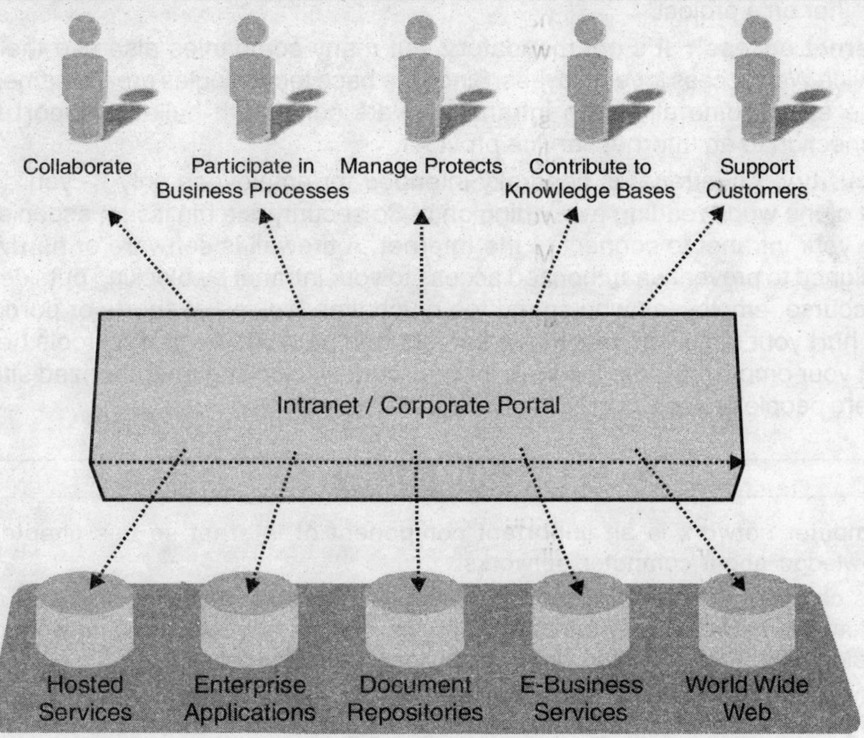

FIGURE 2.10. Intranet.

An intranet's Website looks and act just like other Websites, but has a firewall surrounding it to fend off unauthorized users. Intranets are used to share information. Secure intranets are much less expensive to build and manage than private, proprietary-standard networks.

An intranet is similar in many ways to the Internet — but restricted to users within a single company. (Intranets that extend outside the company are known as extranets.) Intranets use Web servers to distribute information, and users view intranet content with the same Web browsers they would use.

2.3.1. Importance of Intranet

Intranets can solve many of the communications problems that plague businesses of every size. They provide quick access to everything from company handbooks to downloadable insurance forms; employees can find whatever information they need, whether they're at home, at work, or on the road. Companies can save money by putting documents onto an intranet and allowing employees to print them as needed, rather than printing thousands of copies in advance. Even something as simple as a corporate phone directory works better as an intranet document, since employees can instantly make changes and view updates without waiting. Increasingly, both large and small companies are putting mission-critical business practices on their intranets, often realizing huge savings.

2.3.2. The Elements of a Corporate Intranet

- **Group collaboration :** By combining several tools on an intranet, groups can share calendars, take part in "virtual workspaces" that contain public messages and files, and log into private chat areas dedicated to specific business projects. These group collaboration features are ideal if you have employees in several branch offices who need to work together on a project.
- **Internet access :** It´s not mandatory, but many companies also use their intranets to provide Web access for employees. Since the base technologies are the same, it´s generally quite easy. Some all-in-one intranet servers come with built-in support for sharing a connection to an Internet service provider.
- **Security :** An intranet is generally intended for employees only — you don´t want the rest of the world reading everything on it. So security is a big issue, especially if you also use your intranet to connect to the Internet. A firewall is software or hardware, or both, designed to prevent unauthorized access to your intranet by blocking outside connections. Of course, employees who spend too much time looking at sports or pornography sites will hurt your business, too. Proxy servers and network monitoring tools help make sure that your employees use the Web for real work by blocking unauthorized sites or tracking where people browse.

SUMMARY

- Computer network is an important component of internet so this chapter gives basic knowledge about computer networks.
- This chapter covers introduction to computer networks, types of computer networks, use of network, network hardware (switch, router, bridges, repeaters and network card), network topology, internet evolution, internet history, and intranet.
- Networking hardware includes all computers, peripherals, interface cards and other equipment needed to perform data-processing and communications within the network.

- The topology of a network is the geometric representation of the relationship of all the links and linking devices (usually called nodes) to one another. There are four basic topologies possible: mesh, star, bus, and ring.
- The development of the internet is not an end in itself; it is a means by which the people can achieve a broad range of economic and social goals.
- Students can develop their feeling with computer networks and internet by going through this chapter.

SHORT QUESTION ANSWERS

Q. 1. What is Computer Networks?

Ans. A computer network is defined to be an interconnected collection of autonomous computers.

Q. 2. What are different types of Computer Networks?

Ans. The three basic types of networks include :

- Local Area Network (LAN)
- Wide Area Network (WAN)
- Metropolitan Area Network (MAN)

Q. 3. What is Switch?

Ans. A network switch is a small hardware device that joins multiple computers together within one Local Area Network (LAN). Technically, network switches operate at layer two (Data Link Layer) of the OSI model.

Q. 4. What is Repeater?

Ans. Network repeaters regenerate incoming electrical, wireless or optical signals. With physical media like Ethernet or Wi-Fi, data transmissions can only span a limited distance before the quality of the signal degrades. Repeaters attempt to preserve signal integrity and extend the distance over which data can safely travel.

Q. 5. What is Bridges ?

Ans. A bridge is a device that allows you to segment a large network into two smaller, more efficient networks. If you are adding to an older wiring scheme and want the new network to be up-to-date, a bridge can connect the two.

Q. 6. What is Routers?

Ans. A router translates information from one network to another; it is similar to a super intelligent bridge. Routers select the best path to route a message, based on the destination address and origin. The router can direct traffic to prevent head-on collisions, and is smart enough to know when to direct traffic along back roads and shortcuts.

Q. 7. What is Network Topology?

Ans. The term network topology refers to the way in which a network is laid out physically: two or more devices connect to a link; two or more links form a topology.

Q. 8. Define Internet?

Ans. The Internet is a network of networks, linking computers to computers sharing the TCP/IP protocols. Each runs software to provide or "serve" information and/or to access and view information.

Internet is not the only network of networks but it has three components:

1. Network of networks
2. Network with information and resources
3. Users on the network

Q. 9. What do you understand by Internet Congestion?

Ans. Network congestion is the situation in which an increase in data transmissions results in a proportionately smaller increase, or even a reduction, in throughput.

Q. 10. Explain Collaborative Computing & Internet?

Ans. Collaborative computing is a term describing a variety of activities where people interact with one another using desktops, laptops, palmtops, and sophisticated digital cellular phones. As computers are best at handling data and representing information, person-to-person communication is enriched by an ability to share, modify, or collaboratively create data and information.

REVIEW QUESTIONS

1. What do you understand by Computer Networks ?
2. Write short notes on following :
   (i) LAN (ii) WAN (iii) MAN
3. Write short notes on following :
   (i) Workstations (ii) NIC (iii) Repeaters
   (iv) Bridger (v) Routing.
4. Explain different topologies with their disadvantages.
5. Define internet and intranet.
6. Write short notes on following :
   (i) Internet congestion (ii) Business culture on internet
   (iii) Collaborative computing (iv) Intranet.

❑❑❑

3
CHAPTER

Working of Internet

INSIDE THIS CHAPTER

3.1. UNIFORM RESOURCE LOCATOR (URL)

Definition ***URL is an acronym for Uniform Resource Locator and is a reference (an address) to a resource on the Internet.***

The following is an example of a URL which addresses the Java Web site hosted by Sun Microsystems:

```
http  :  //java.sun.com
```

Protocol Identifier (http) — Resource Name (//java.sun.com)

As in the previous diagram, a URL has two main components :

- Protocol identifier
- Resource name

Note that the protocol identifier and the resource name are separated by a colon and two forward slashes. The protocol identifier indicates the name of the protocol to be used to fetch the resource. The example uses the Hypertext Transfer Protocol (HTTP), which is typically used to serve up hypertext documents. HTTP is just one of many different protocols used to access different types of resources on the net. Other protocols include File Transfer Protocol (FTP), Gopher, File, and News.

The resource name is the complete address to the resource. The format of the resource name depends entirely on the protocol used, but for many protocols, including HTTP, the resource name contains one or more of the components listed in the following table :

Host Name	The name of the machine on which the resource lives.
Filename	The pathname to the file on the machine.
Port Number	The port number to which to connect (typically optional).
Reference	A reference to a named anchor within a resource that usually identifies a specific location within a file (typically optional).

For many protocols, the host name and the filename are required, while the port number and reference are optional. For example, the resource name for an HTTP URL must specify a server on the network (Host Name) and the path to the document on that machine (Filename); it also can specify a port number and a reference. In the URL for the Java Web site java.sun.com is the host name and the trailing slash is shorthand for the file named/index.html.

3.2. INTERNET ADDRESS (IP ADDRESS) [MDU Dec. 2006]

Definition

Every machine on the Internet has a unique identifying number, called an IP Address. The IP stands for Internet Protocol, which is the language that computers use to communicate over the Internet.

A protocol is the predefined way that someone who wants to use a service talks with that service. The "someone" could be a person, but more often it is a computer program like a Web browser.

A typical IP address looks like this:

216.27.61.137

To make it easier for us humans to remember, IP addresses are normally expressed in decimal format as a dotted decimal number like the one above. But computers communicate in binary form. Look at the same IP address in binary.

11011000.00011011.00111101.10001001

The four numbers in an IP address are called octets, because they each have eight positions when viewed in binary form. If we add all the positions together, we get 32, which is why IP addresses are considered 32-bit numbers. Since each of the eight positions can have two different states (1 or zero), the total number of possible combinations per octet is 2^8 or 256. So each octet can contain any value between zero and 255. Combine the four octets and you get 2^{32} or a possible 4,294,967,296 unique values.

TIPS

Out of the almost 4.3 billion possible combinations, certain values are restricted from use as typical IP addresses. For example, the IP address 0.0.0.0 is reserved for the default network and the address 255.255.255.255 is used for broadcasts.

The octets serve a purpose other than simply separating the numbers. They are used to create classes of IP addresses that can be assigned to a particular business, government or other entity based on size and need. The octets are split into two sections: Net and Host. The Net section always contains the first octet. It is used to identify the network that a computer belongs to. Host (sometimes referred to as Node) identifies the actual computer on the network. The Host section always contains the last octet. There are five IP classes plus certain special addresses

3.3. DOMAIN NAMES

Definition

A name that identifies one or more IP addresses is called domain name. For example, the domain name microsoft.com represents about a dozen IP addresses.

Domain names are used in URLs to identify particular Web pages. For example, in the URL *http://www.kiet.edu/index.html,* the domain name is kiet.edu.

Every domain name has a suffix that indicates which top level domain it belongs to. There are only a limited number of such domains. For example:

- gov - Government agencies
- edu - Educational institutions
- org - Organizations (nonprofit)
- mil - Military
- com - Commercial business
- net - Network organizations
- ca - Canada
- th - Thailand
- in - India

TIPS

Because the Internet is based on IP addresses, not domain names, every Web server requires a Domain Name System (DNS) server to translate domain names into IP addresses.

3.3.1. Domain Name System

[MDU May 2006]

Definition

DNS stands for two things: Domain Name Service and Domain Name Servers. One acronym defines the protocol, the other defines the machines that provide the service. The job that DNS performs is very simple: it takes the IP addresses that computers connected to the Internet use to communicate with each other and it maps them to host names.

Human beings tend to have a difficult time remembering long strings of seemingly arbitrary numbers. The way that our brains work, it's difficult to make information like that stick. And that is where DNS comes in. It allows us to substitute words or phrases for those strings of numbers. Words are a lot easier for people to remember than numbers, especially when they can be tied to a specific idea that is linked to the website.

DNS translates IP addresses into hostnames and back again. The hostnames are for the benefit of human end users. The IP addresses are the only essential thing, as far as the computers are concerned.

3.4. ISP (INTERNET SERVICE PROVIDERS) [MDU May 2006, 2008]

For networks connected to the global Internet, an organization obtains network number from the communication company that supplies internet connections. Such companies are called Internet Service Provider (ISPs).

Definition ***ISP is an organization or business offering public access to the Internet. It is our gateway, to the Net.***

You have to subscribe to a provider for your Internet connection. You use your computer and modem to access the provider's system and the provider handles the rest of the details of connecting you to the Internet.

3.4.1. Types of ISP

There are many types of Internet providers. We can, for instance, choose one of the big commercial on-line service providers. The primary business of an **ISP** is hooking people to the Internet by giving an Internet account to subscribers, and providing them with two different kinds of access:

1. Shell access and
2. SLIP/PPP access.

Most **ISP** offer both kinds of access, some offer both with a single account, and others require that you choose one or the other. Once you register, your provider will give you a user name (called a user ID), a password, and a phone number to dial. To establish the Internet connection, you have your communications program dial the number. You then log in using your particular user ID and password.

At present it is **VSNL** (Videsh Sanchar Nigam Limited) which is dominating the Internet scene in India through its GIAS (Gateway Internet Access Service). The other service providers in India are **MTNL** (Mahanagar Telephone Nigam Limited), Mantra-online, Airtel, Reliance Communications and Satyam-online.

3.5. MODEMS

A modem is a Modulator/Demodulator; it connects a terminal/computer (DTE) to the Voice Channel (dial-up line).

Definition ***"A modem is a device or program that enables a computer to transmit data over, for example, telephone or cable lines. Computer information is stored digitally, whereas information transmitted over telephone lines is transmitted in the form of analog waves. A modem converts between these two forms".***

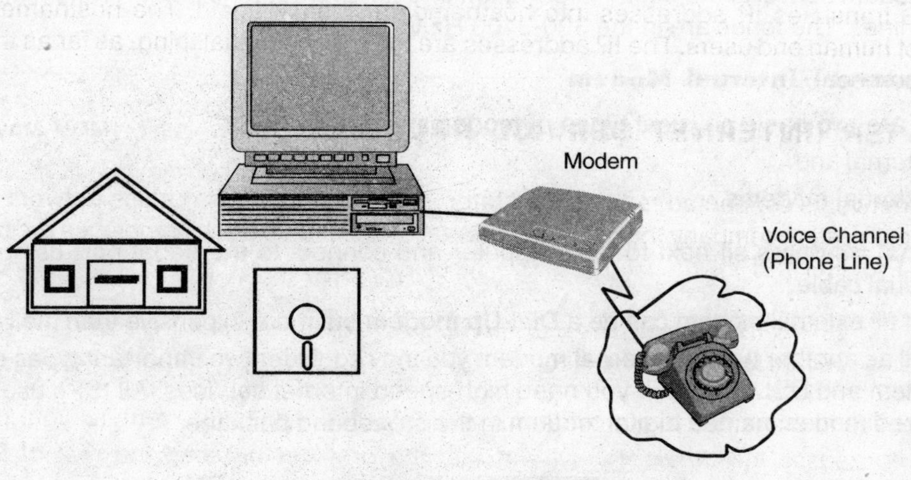

FIGURE 3.1. Modem.

A modem can established two type of connections as follows :

1. Digital Connection

The connection between the modem and terminal/computer is a digital connection. A basic connection consists of a Transmit Data (TXD) line, a Receive Data (RXD) line and many hardware hand-shaking control lines.

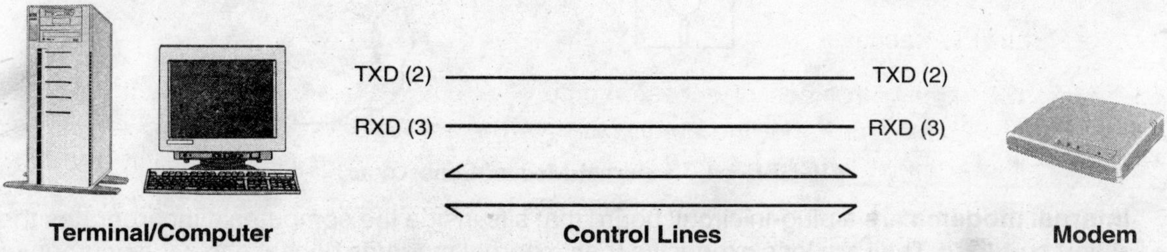

FIGURE 3.2. Digital Connection.

The control lines determine: whose turn it is to talk (modem or terminal), if the terminal/computer is turned on, if the modem is turned on, if there is a connection to another modem, etc.

2. Analog Connection

The connection between the modem and outside world (phone line) is an analog connection. The Voice Channel has a bandwidth of 0-4 kHz but only 300-3400 Hz is usable for data communications.

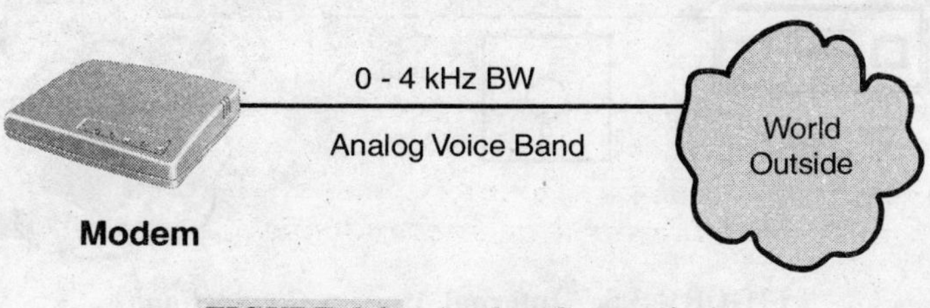

FIGURE 3.3. Analog Connection.

The modem converts the digital information into tones (frequencies) for transmitting through the phone lines. The tones are in the 300-3400 Hz Voice Band.

3.5.1. External/Internal Modem

There are two basic physical types of modems :

1. Internal and
2. External modems.

External modems sit next to the computer and connect to the serial port using a straight through serial cable.

First of all external modem can be a **Dial-Up modem** but more expensive than the internal.

As well as another type of external modem you may consider two important types of modem : cable modem and DSL modem if you need high-speed internet services. All ISPs usually supply a specialized modem named digital modem in the broadband package.

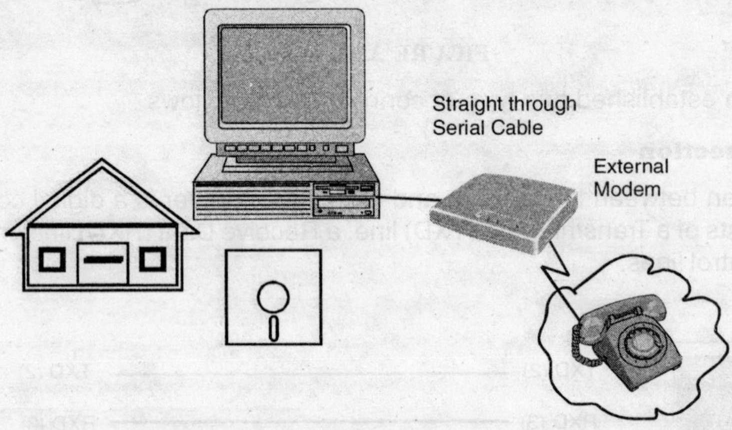

FIGURE 3.4. External Modem Connections.

Internal modems are a plug-in circuit board that sits inside the computer. It incorporates the serial port on-board. They are less expensive than external modems because they do not require a case, power supply and serial cable. They appear to the communication programs as if they were an external modem for all intensive purposes.

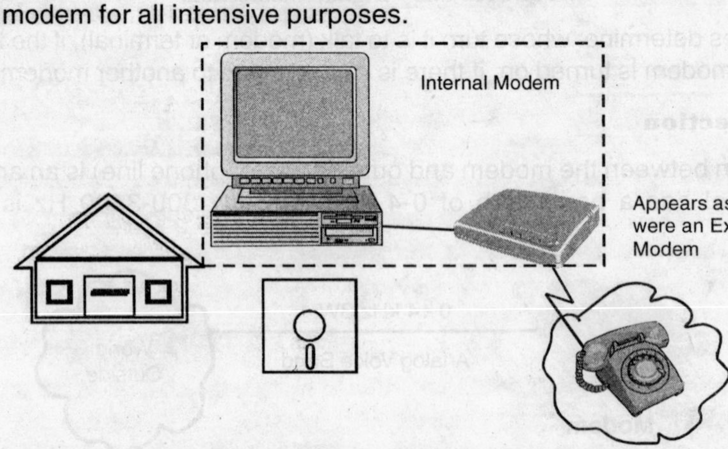

FIGURE 3.5. Internal Modem Connection.

Internal computer modems are used with Dial-Up Internet connection through a couple of RJ-11 connection. Also called copper telephone lines, the RJ-11 allows computer to receive and send data as explained above.

3.6. INTRODUCTION TO CONNECTIVITY [MDU May 2005, 2006]

To connect our computer to the Internet, we have to do certain formalities and we have to take certain decision, like :

1. What type of connection we want ?
2. What will we use the service for?
3. Do we need significant bandwidth to send information, or will you use it primarily to surf the Internet, where upstream bandwidth is not as critical?
4. How much money we want to spend ?

3.6.1. Types of Internet Account

To connect to the net, we run a communication program to dial the number of an Internet service provider. When the remote system answers the phone, your modem talks to the other system and a connection is established.

There are two possible types of connections :

1. Shell Account
2. PPP (point to point protocol) account.

3.6.2. PPP Account

The first type of Internet account is called PPP account :

1. This type of account lets you connect your computer right onto the Internet.
2. When you have a PPP account, your computer is a full-fledged Internet host during the time you are connected ?
3. With this type of account, your own computer does most of the work and the remote computer acts mainly as a conduit into the net.
4. For the duration of your Internet session, your computer is a part of the net, just like a computer directly connected with a cable. The only difference is your computer is connected over a telephone line.
5. As far as the rest of the net is concerned, you are on the net and other Internet computers can communicate with your computers directly.
6. When you use a PPP connection, your computer temporarily becomes an Internet host ?
7. This means the client program you use to access the various Internet services run on your computer.
8. The best thing about a PPP connection is you can use as many clients as you want at the same time.
9. The best way to think about a PPP connection is as a tunnel into the net.
10. You run whatever client programs you want on your computer and when they need to send or receive information, they use the tunnel.
11. The nature of the tunnel is such that it will support as many clients as you want to use. Of course, if you make too many demands on the system at the same time, the traffic in the tunnel will increase and everything will slow down.

3.6.3. Shell Account

With this type of account we actually do your work on the remote computer. Your computer is not on the net itself, rather your computer emulates a terminal, which allows you to work directly with the remote host.

When we use shell account, the only program that runs on your computer is the terminal emulates. All the Internet clients actually run on the remote computer. The nature of a terminal emulator is that it only supports a text-based interface, not a graphical interface.

The client programs run on the remote host not on your computer. All your computer does is emulate a terminal. This means you do not run programs on your computer, so you can't take advantage of the characteristics of your computer; you are limited to the capabilities of the terminal your computer is emulating.

The main three **disadvantages** to a shell account :

1. The interface is text based not graphical.
2. Your computer is not on the Internet.
3. You are usually limited to running one client at a time.

3.7. MODES OF CONNECTING TO INTERNET

If we are looking for a better Internet connection then we must think about high-speed Internet connections, such as cable, ISDN and DSL but need a little clarification as to what each of these different types of connections are and what they do.

3.7.1. Telephone Line Internet Connectivity (Dial-up connection)

The most basic type of Internet connection is called a dial-up connection. This connection is made through a modem (the communication mechanism in computers) that uses a telephone line to connect to the Internet. The modem must dial the telephone every time it wants to connect to the Internet, hence the name **dial-up connection**. Most people connect to the Internet by using a modem and phone line to dial in to a PPP account on an Internet provider's computer. Most ISP's support modems at speeds 14.4Kbps, 28.2Kbps, and 56Kbps.

3.7.2. ISDN - Dial-Up Connection

The second type of dial-up connection is through an ISDN (Integrated Services Digital Network). This connection is a high-speed dial-up and requires a special type of dedicated *digital* telephone line. This type of connection improves speed because the signals received are already in a digital format, but it is costly. For a while dial-up connections were the only way to access the Internet. Now, however, broadband connections are available in many parts of the country as telephone and cable companies install new lines to accommodate Internet connections.

3.7.3. Broadband Connections

Broadband is a high-speed Internet connection that makes surfing the Web more enjoyable. It also easily accommodates the video, audio, or complex graphics that are becoming commonplace on the Internet. Broadband is a type of data transmission in which a single medium or wire can carry several channels or communication paths at once. So broadband cable connections are able to transmit both television signals and Internet data at the same time. Broadband telephone connections like DSL are able to transmit both voice and data over the same line at the same time.

Broadband connections are always on. That means if your computer is on, you will be connected to the Internet with no dial-up involved. For most average users, a broadband connection will mean accessing the Internet by either a cable modem provided by their local cable company or a DSL modem and DSL telephone line provided by their local telephone service provider.

3.7.4. Internet Through Cable

Internet over cable is a technology and service that uses the existing cable TV networks (made up of coaxial cable lines that bring television signals to TV) to bring Internet data to PC or TV at a very high speed. The television and the Internet transmission take place simultaneously on the same cable but at different frequencies. This allows the user to view TV and access Internet at the same time.

There are only two drawbacks to cable:

- First, it is a shared connection, meaning you share the "pipeline" with your neighbours. That doesn't mean that your neighbours will know what you are doing on the Internet.
- The second drawback to cable is also minor. Because cable modem connections are always on, they, like DSL connections, make you more vulnerable to hacking and security breaches. For this reason, many cable companies are now providing their customers firewall software to help protect their security.

3.7.5. Internet Connectivity through Leased Line

Leased lines are dedicated circuits provided by Basic Service Providers (BSPs), which provide permanent connectivity to the Internet. Leased lines provide the last mile access from the user premises to the ISP. They provide permanent connection as compared to the temporary connectivity through dialup access. The quality of the connection is far superior to what is normally available through dialup, thanks to digital signaling, less noise, fewer exchanges etc.

Leased lines provides a scalable access method, important particularly for organizations with large user groups, including corporate, banks and financial institutions, educational and R&D organizations, government, military etc. Starting typically with 64 kbps, it is possible to deploy a scalable architecture, with multiples of E1 (2 MBPS) pipes, providing the necessary bandwidth. In fact, leased access becomes a must for large organizations in most situations.

Advantages

It provides permanent, reliable, high-speed connectivity as compared to the temporary connectivity of dial up access. The quality of the connection is far superior to what is normally available through dialup, because of the digital signaling, less noise, fewer exchanges etc.

Disadvantages

Leased bandwidth prices are quite high, compared to dialup bandwidth of comparable size. Entry level annual port prices are also high at present, so that this access method is only feasible beyond a fairly high threshold level.

Permanent connectivity to the Net exposes the organization to a variety of threats including hacking, malicious code including active vandals, viruses, Trojan Horses, macros, denial of service attacks etc.

3.8. WORKING OF INTERNET [MDU May 2005, 2006]

The Internet's growth has become explosive and it seems impossible to escape the bombardment of *www.com*'s seen constantly on television, heard on radio, and seen in magazines. Because the Internet has become such a large part of our lives, a good understanding is needed to use this new tool most effectively. It is important for the students to understand the working of internet. The working of internet can be seen in following steps :

1. Connection to internet.
2. Communication with other computers.
3. Networking infrastructure.
4. Internet infrastructure.
5. The Internet routing hierarchy.
6. Domain name and address resolution.

3.8.1. Connect to Internet

The picture below illustrates two computers connected to the Internet; your computer with IP address 1.2.3.4 and another computer with IP address 5.6.7.8. The Internet is represented as an abstract object in-between.

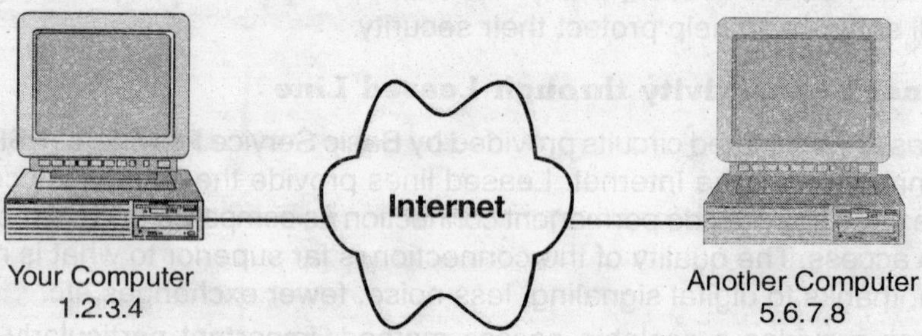

FIGURE 3.6. Connection to Internet.

If you connect to the Internet through an Internet Service Provider (ISP), you are usually assigned a temporary IP address for the duration of your dial-in session. If you connect to the Internet from a Local Area Network (LAN) your computer might have a permanent IP address or it might obtain a temporary one from a DHCP (Dynamic Host Configuration Protocol) server. In any case, if you are connected to the Internet, your computer has a unique IP address.

3.8.2. Communication With Other Computers

So your computer is connected to the Internet and has a unique address. How does it 'talk' to other computers connected to the Internet? An example should serve here: Let's say your IP address is 1.2.3.4 and you want to send a message to the computer 5.6.7.8. The message you want to send is "Hello computer 5.6.7.8!". Obviously, the message must be transmitted over whatever kind of wire connects your computer to the Internet. Let's say you've dialed into your ISP from home and the message must be transmitted over the phone line. Therefore the message must be translated from alphabetic text into electronic signals, transmitted over the Internet, then translated back into alphabetic text. How is this accomplished? Through the use of a **protocol**

stack. Every computer needs one to communicate on the Internet and it is usually built into the computer's operating system (i.e. Windows, Unix, etc.). The protocol stack used on the Internet is referred to as the TCP/IP protocol stack because of the two major communication protocols used. The TCP/IP stack looks like this :

S. No.	Protocol Layer	Comments
1.	Application Protocols Layer	Protocols specific to applications such as WWW, e-mail, FTP, etc.
2.	Transmission Control Protocol Layer	TCP directs packets to a specific application on a computer using a port number.
3.	Internet Protocol Layer	IP directs packets to a specific computer using an IP address.
4.	Hardware Layer (Physical Layer)	Converts binary packet data to network signals and back. (E.g. ethernet network card, modem for phone lines, etc.)

If we were to follow the path that the message "Hello computer 5.6.7.8!" took from our computer to the computer with IP address 5.6.7.8, it would happen something like this :

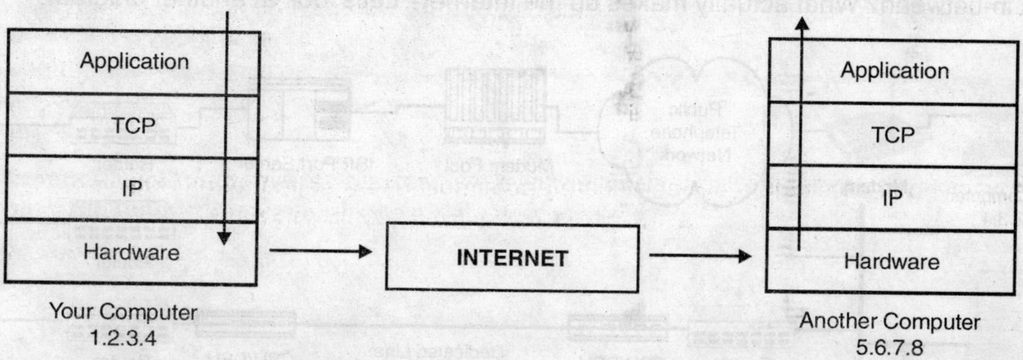

FIGURE 3.7. Flow of Message on Internet.

1. The message would start at the top of the protocol stack on your computer and work it's way downward.
2. If the message to be sent is long, each stack layer that the message passes through may break the message up into smaller chunks of data. This is because data sent over the Internet (and most computer networks) are sent in manageable chunks. On the Internet, these chunks of data are known as **packets**.
3. The packets would go through the Application Layer and continue to the TCP layer. Each packet is assigned a **port number**. Ports will be explained later, but suffice to say that many programs may be using the TCP/IP stack and sending messages. We need to know which program on the destination computer needs to receive the message because it will be listening on a specific port.
4. After going through the TCP layer, the packets proceed to the IP layer. This is where each packet receives it's destination address, 5.6.7.8.

5. Now that our message packets have a port number and an IP address, they are ready to be sent over the Internet. The hardware layer takes care of turning our packets containing the alphabetic text of our message into electronic signals and transmitting them over the phone line.
6. On the other end of the phone line your ISP has a direct connection to the Internet. The ISPs **router** examines the destination address in each packet and determines where to send it. Often, the packet's next stop is another router. More on routers and Internet infrastructure later.
7. Eventually, the packets reach computer 5.6.7.8. Here, the packets start at the bottom of the destination computer's TCP/IP stack and work upwards.
8. As the packets go upwards through the stack, all routing data that the sending computer's stack added (such as IP address and port number) is stripped from the packets.
9. When the data reaches the top of the stack, the packets have been re-assembled into their original form, "Hello computer 5.6.7.8!"

3.8.3. Networking Infrastructure

So now you know how packets travel from one computer to another over the Internet. But what's in-between? What actually makes up the Internet? Let's look at another diagram:

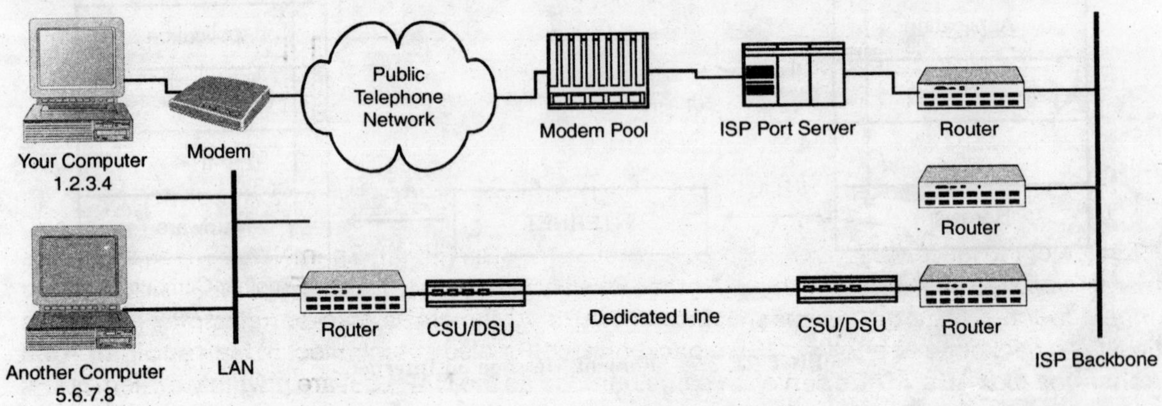

FIGURE 3.8. Packet Traveling Through Internet.

Here we see Fig. 3.8 redrawn with more detail. The physical connection through the phone network to the Internet Service Provider might have been easy to guess, but beyond that might bear some explanation.

The ISP maintains a pool of modems for their dial-in customers. This is managed by some form of computer (usually a dedicated one) which controls data flow from the modem pool to a backbone or dedicated line router. This setup may be referred to as a port server, as it 'serves' access to the network. Billing and usage information is usually collected here as well.

After your packets traverse the phone network and your ISP's local equipment, they are routed onto the ISP's backbone or a backbone the ISP buys bandwidth from. From here the packets will usually journey through several routers and over several backbones, dedicated lines, and other networks until they find their destination, the computer with address 5.6.7.8.

3.8.4. Internet Infrastructure

The Internet backbone is made up of many large networks which interconnect with each other. These large networks are known as **Network Service Providers** or **NSPs**.

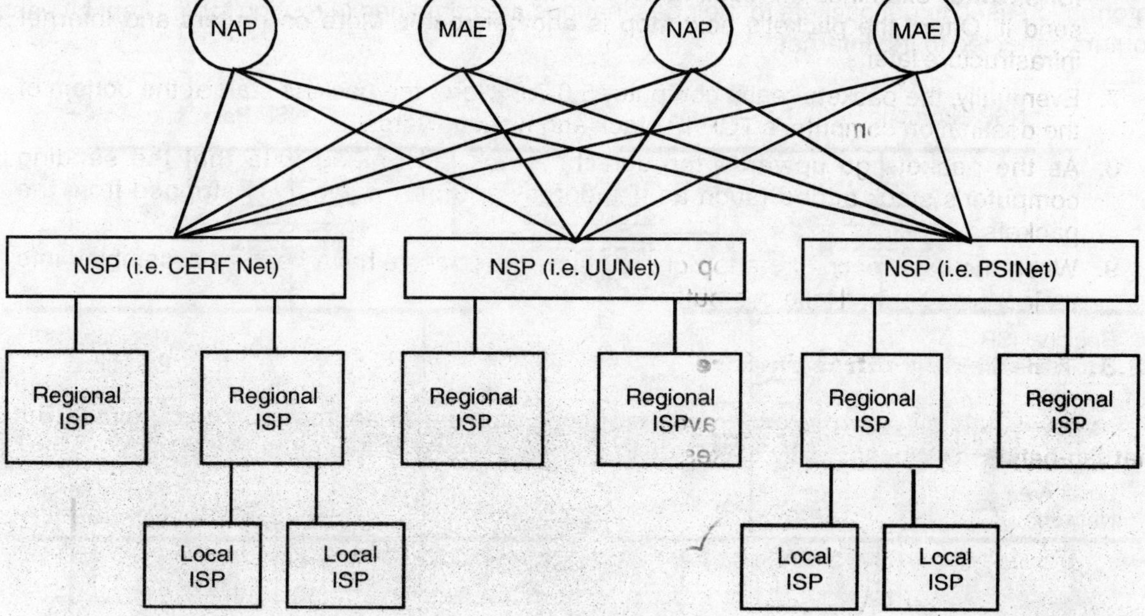

FIGURE 3.9. Internet Infrastructure.

Some of the large NSPs are UUNet, CerfNet, IBM, BBN Planet, SprintNet, PSINet, as well as others. These networks **peer** with each other to exchange packet traffic. Each NSP is required to connect to three **Network Access Points** or **NAP**s. At the NAPs, packet traffic may jump from one NSP's backbone to another NSP's backbone. NSPs also interconnect at **Metropolitan Area Exchanges** or **MAE**s. MAEs serve the same purpose as the NAPs but are privately owned. NAPs were the original Internet interconnect points. Both NAPs and MAEs are referred to as Internet Exchange Points or **IX**s. NSPs also sell bandwidth to smaller networks, such as ISPs and smaller bandwidth providers. Below is a picture showing this hierarchical infrastructure.

TIPS

This is not a true representation of an actual piece of the Internet. Fig. 3.8 is only meant to demonstrate how the NSPs could interconnect with each other and smaller ISPs. None of the physical network components are shown in Fig. 3.8 as they are in Fig. 3.7. This is because a single NSP's backbone infrastructure is a complex drawing by itself. Most NSPs publish maps of their network infrastructure on their websites and can be found easily. To draw an actual map of the Internet would be nearly impossible due to it's size, complexity, and ever changing structure.

3.8.5. The Internet Routing Hierarchy

So how do packets find their way across the Internet ? Does every computer connected to the Internet know where the other computers are ? Do packets simply get 'broadcast' to every computer on the Internet? The answer to both the preceding questions is 'no'. No computer knows where any of the other computers are, and packets do not get sent to every computer. The information used to get packets to their destinations are contained in routing tables kept by each router connected to the Internet.

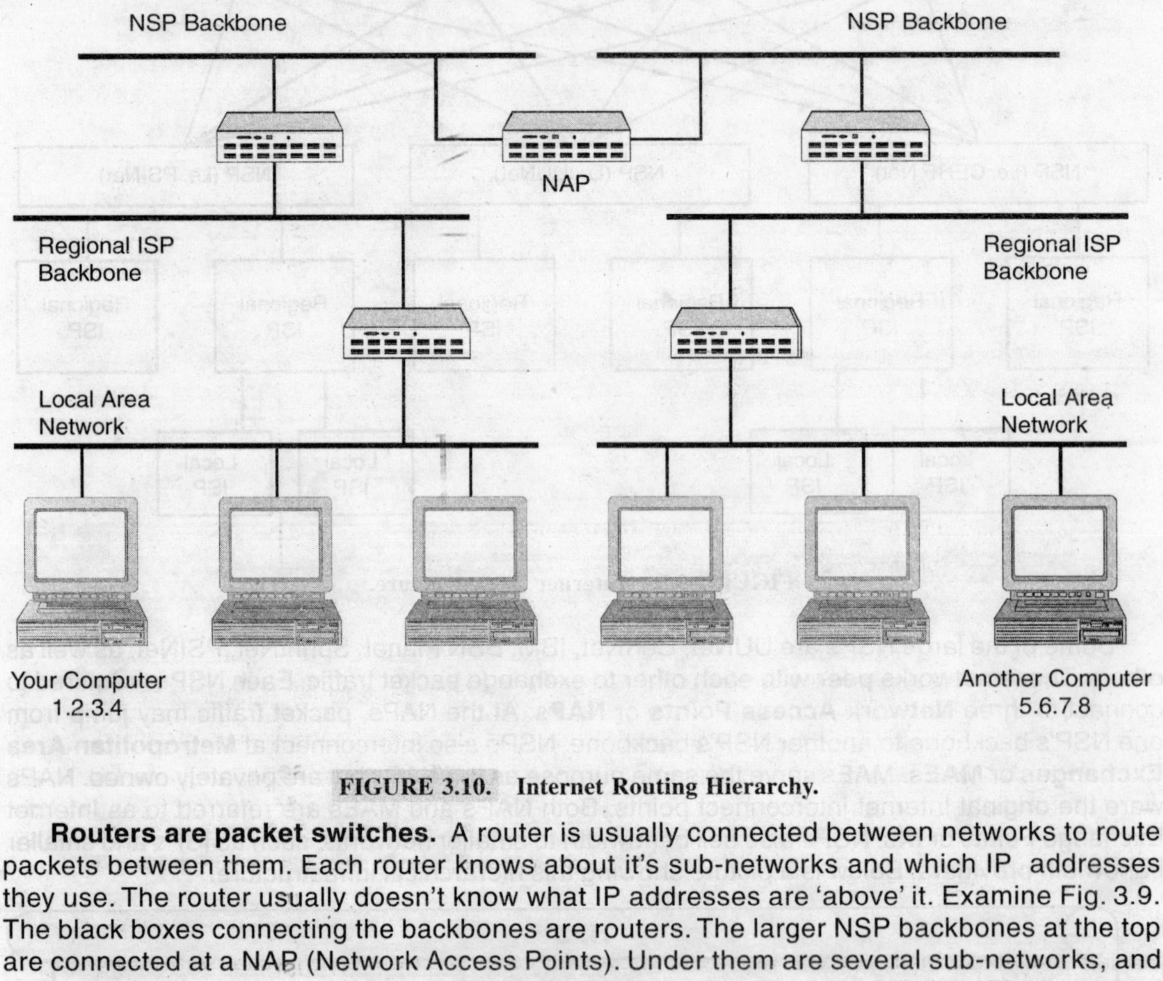

FIGURE 3.10. Internet Routing Hierarchy.

Routers are packet switches. A router is usually connected between networks to route packets between them. Each router knows about it's sub-networks and which IP addresses they use. The router usually doesn't know what IP addresses are 'above' it. Examine Fig. 3.9. The black boxes connecting the backbones are routers. The larger NSP backbones at the top are connected at a NAP (Network Access Points). Under them are several sub-networks, and under them, more sub-networks. At the bottom are two local area networks with computers attached.

When a packet arrives at a router, the router examines the IP address put there by the IP protocol layer on the originating computer. The router checks it's routing table. If the network containing the IP address is found, the packet is sent to that network. If the network containing the IP address is not found, then the router sends the packet on a default route, usually up the backbone hierarchy to the next router. Hopefully the next router will know where to send the packet. If it does not, again the packet is routed upwards until it reaches a NSP backbone. The routers connected to the NSP backbones hold the largest routing tables and here the packet will be routed to the correct backbone, where it will begin its journey 'downward' through smaller and smaller networks until it finds its destination.

3.8.6. Domain Names and Address Resolution

But what if you don't know the IP address of the computer you want to connect to? What if the you need to access a web server referred to as *www.anothercomputer.com*? How does your web browser know where on the Internet this computer lives? The answer to all these questions is the **Domain Name Service** or **DNS**. The DNS is a distributed database which keeps track of computer's names and their corresponding IP addresses on the Internet.

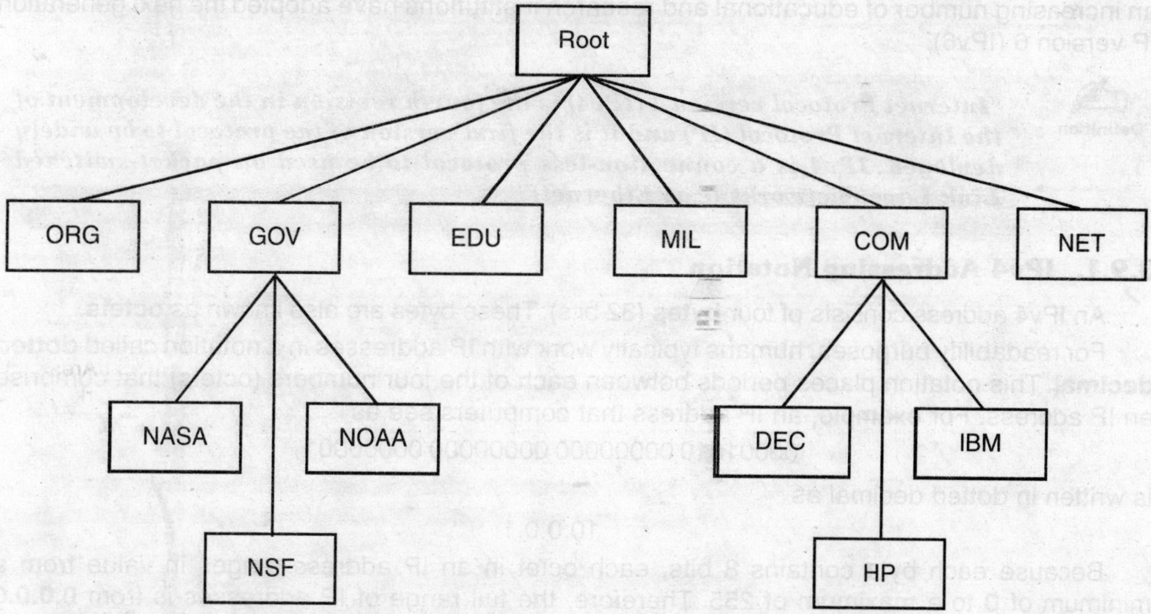

FIGURE 3.11. Domain Names and Address Resolution.

Many computers connected to the Internet host part of the DNS database and the software that allows others to access it. These computers are known as DNS servers. No DNS server contains the entire database; they only contain a subset of it. If a DNS server does not contain the domain name requested by another computer, the DNS server re-directs the requesting computer to another DNS server.

The Domain Name Service is structured as a hierarchy similar to the IP routing hierarchy. The computer requesting a name resolution will be re-directed 'up' the hierarchy until a DNS server is found that can resolve the domain name in the request. Figure 3.6 illustrates a portion of the hierarchy. At the top of the tree are the domain roots. Some of the older, more common domains are seen near the top. What is not shown are the multitude of DNS servers around the world which form the rest of the hierarchy.

When an Internet connection is setup (e.g. for a LAN or Dial-Up Networking in Windows), one primary and one or more secondary DNS servers are usually specified as part of the installation. This way, any Internet applications that need domain name resolution will be able to function correctly. For example, when you enter a web address into your web browser, the browser first connects to your primary DNS server. After obtaining the IP address for the domain name you entered, the browser then connects to the target computer and requests the web page you wanted.

3.9. IPv4 [MDU May 2006, 2008]

Internet Protocol (IP) technology was developed in the 1970s to support some of the first research computer networks. Today, IP has become a worldwide standard for home and business networking as well. Our network routers, Web browsers, email programs, instant messaging software - all rely on IP or other network protocols layered on top of IP. Two versions of IP technology exist today. Essentially all home computer networks use IP version 4 (IPv4), but an increasing number of educational and research institutions have adopted the next generation IP version 6 (IPv6).

Definition

"Internet Protocol version 4 (IPv4) is the fourth revision in the development of the Internet Protocol (IP) and it is the first version of the protocol to be widely deployed. IPv4 is a connection-less protocol to be used on packet-switched Link Layer networks (e.g., Ethernet)".

3.9.1. IPv4 Addressing Notation

An IPv4 address consists of four **bytes** (32 bits). These bytes are also known as **octets.**

For readability purposes, humans typically work with IP addresses in a notation called **dotted decimal**. This notation places periods between each of the four numbers (octets) that comprise an IP address. For example, an IP address that computers see as

00001010 00000000 00000000 00000001

is written in dotted decimal as

10.0.0.1

Because each byte contains 8 bits, each octet in an IP address ranges in value from a minimum of 0 to a maximum of 255. Therefore, the full range of IP addresses is from **0.0.0.0** through **255.255.255.255**. That represents a total of 4,294,967,296 possible IP addresses.

3.9.2. IPv4 Header

The header (Internet Protocol version 4) is describing as follows:

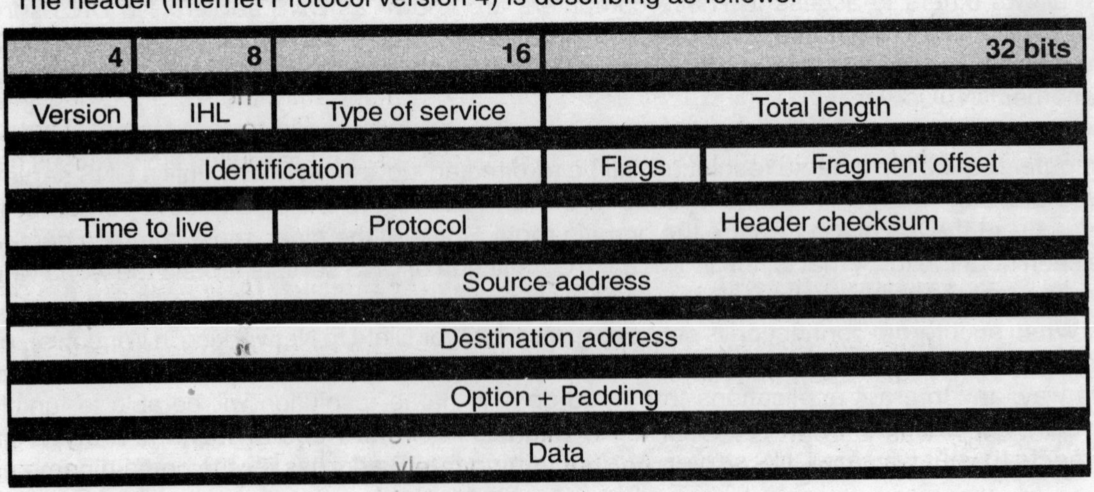

FIGURE 3.12. IPv4 Header.

- **Version :** The version of IP currently used.
- **IP Header Length (IHL) :** Datagram header length. Points to the beginning of the data. The minimum value for a correct header is 5.
- **Type-of-Service :** Indicates the quality of service desired by specifying how an upper-layer protocol would like a current datagram to be handled, and assigns datagrams various levels of importance. This field is used for the assignment of Precedence, Delay, Throughput and Reliability.
- **Total Length :** Specifies the length, in bytes, of the entire IP packet, including the data and header. The maximum length could be specified by this field is 65,535 bytes. Typically, hosts are prepared to accept datagrams up to 576 bytes.
- **Identification :** Contains an integer that identifies the current datagram. This field is assigned by sender to help receiver to assemble the datagram fragments.
- **Flags :** Consists of a 3-bit field of which the two low-order (least-significant) bits control fragmentation. The low-order bit specifies whether the packet can be fragmented. The middle bit specifies whether the packet is the last fragment in a series of fragmented packets. The third or high-order bit is not used.
- **Fragment Offset :** This 13 bits field indicates the position of the fragment's data relative to the beginning of the data in the original datagram, which allows the destination IP process to properly reconstruct the original datagram.
- **Time-to-Live :** It is a counter that gradually decrements down to zero, at which point the datagram is discarded. This keeps packets from looping endlessly.
- **Protocol :** Indicates which upper-layer protocol receives incoming packets after IP processing is complete.
- **Header Checksum :** Helps ensure IP header integrity. Since some header fields change, e.g., Time To Live, this is recomputed and verified at each point that the Internet header is processed.
- **Source Address :** Specifies the sending node.
- **Destination Address :** Specifies the receiving node.
- **Options :** Allows IP to support various options, such as security.

3.10. IPv6 [MDU May 2006, Dec. 2007, May 2008]

Most of today's internet uses IPv4, which is now nearly twenty years old. IPv4 was efficient but in spite of that it is beginning to have problems. Most importantly, there is a growing shortage of IPv4 addresses, which are needed by all new machines added to the Internet. The routing tables of Internet backbone routers are becoming larger. A separate routing table entry is needed for each network resulting in a large number of routing table entries.

Definition

The Internet Engineering Task Force (IETF) has designated IPv6 as the successor to version 4 for general use on the Internet. It significantly increases the size of the address space used to identify communication endpoints in the Internet, thereby allowing it to continue its tremendous growth rate. IPv6 is also known as IPng (IP Next Generation).

Security was also an issue for IPv4. Although there are lots of ways of encrypting IPv4 traffic, such as using the IPSec protocol, but unfortunately all of the IPv4 encryption methods are proprietary and no real standard encryption methods exist.

3.10.1. IPv6 Addressing Notation

IP addresses change significantly with IPv6. IPv6 addresses are 16 bytes (128 bits) long rather than four bytes (32 bits). This larger size means that IPv6 supports more than

300,000,000,000,000,000,000,000,000,000,000,000,000

In the coming years, as an increasing number of cell phones, PDAs, and other consumer electronics expand their networking capability, the smaller IPv4 address space will likely run out and IPv6 address become necessary.

IPv6 addresses are generally written in the following form :

hhhh:hhhh:hhhh:hhhh:hhhh:hhhh:hhhh:hhhh

In this **full notation**, pairs of IPv6 bytes are separated by a colon and each byte in turns is represented as a pair of hexadecimal numbers, like in the following example :

E3D7:0000:0000:0000:51F4:9BC8:C0A8:6420

As shown above, IPv6 addresses commonly contain many bytes with a zero value. **Shorthand notation** in IPv6 removes these values from the text representation (though the bytes are still present in the actual network address) as follows :

E3D7::51F4:9BC8:C0A8:6420

Finally, many IPv6 addresses are extensions of IPv4 addresses. In these cases, the rightmost four bytes of an IPv6 address (the rightmost two byte pairs) may be rewritten in the IPv4 notation. Converting the above example to **mixed notation** yields

E3D7::51F4:9BC8:192.168.100.32

IPv6 addresses may be written in any of the full, shorthand or mixed notation illustrated above.

3.10.2. IPv6 Header

IPv6 header is as follows:

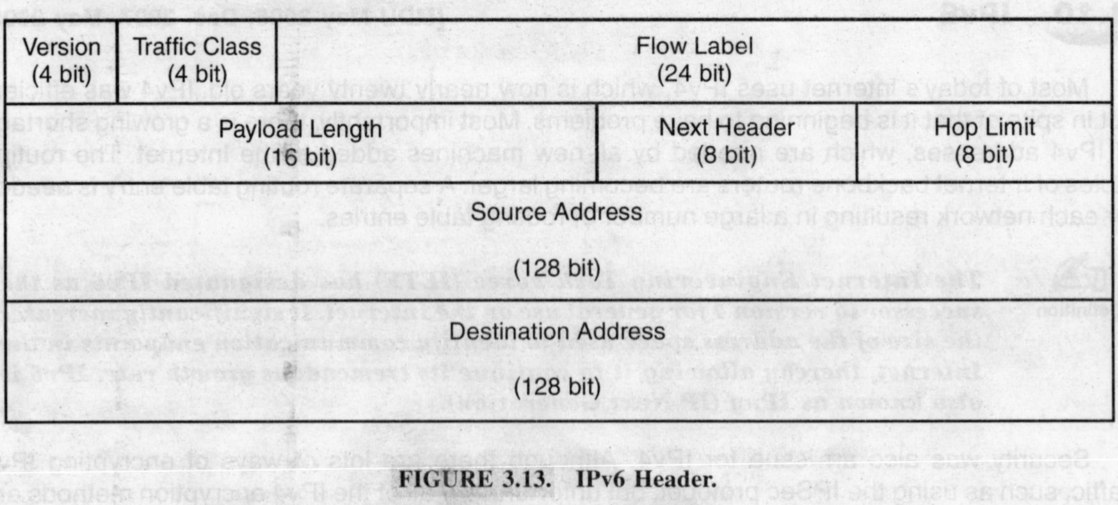

FIGURE 3.13. IPv6 Header.

IPv6 header contains the following things :

- **Version :** This field contains the version of the IP used in the packet. It is of 4-bit in IP version 6.
- **Traffic class :** This is an 8-bit field determining the packet priority. Priority values subdivide into ranges: traffic where the source provides congestion control and non-congestion control traffic.
- **Flow label :** This 20 bits specifies the QoS management. Originally created for giving real-time applications special service, but currently unused.
- **Payload length :** This 16 bits determines the payload length in bytes. When cleared to zero, the option is a "Jumbo payload" (hop-by-hop) ?
- **Next header :** This 8-bit field specifies the next encapsulated protocol. The values are compatible with those specified for the IPv4 protocol field.
- **Hop limit :** This is an 8-bit field newly introduced in IPv6. It replaces the time to live field of IPv4.
- **Source Address :** This 128 bits field determines the logical address of the host that is sending the packet.
- **Destination Address :** This 128 bits field determines the logical address of the host that is receiving the packet.
- The payload can have a size of up to 64KB in standard mode, or larger with a "jumbo payload" option.

3.10.3. Difference between IPv4 and IPv6 [MDU May 2006]

The difference between IPv4 and IPv6 can be understood form following comparison table :

S. No.	IPv4	IPv6
1.	Source and destination addresses are 32 bits (4 bytes) in length.	Source and destination addresses are 128 bits (16 bytes) in length.
2.	IPSec support is optional.	IPSec support is required.
3.	IPv4 header does not identify packet flow for QoS handling by routers.	IPv6 header contains Flow Label field, which identifies packet flow for QoS handling by router.
4.	Both routers and the sending host fragment packets.	Only the sending host fragments packets; routers do not.
5.	Header includes a checksum.	Header does not include a checksum.
6.	IPSec support is optional.	All optional data is moved to IPv6 extension headers.
7.	IPv4 header does not identify packet flow for QoS handling by routers.	Multicast Neighbor Solicitation messages resolve IP addresses to link-layer addresses.
8.	Both routers and the sending host fragment packets.	Multicast Listener Discovery (MLD) messages manage membership in local subnet groups.

S. No.	IPv4	IPv6
9.	Header includes a checksum.	ICMPv6 Router Solicitation and Router Advertisement messages are used to determine the IP address of the best default gateway, and they are required.
10.	IPSec support is optional.	IPv6 uses a link-local scope all-nodes multicast address.
11.	IPv4 header does not identify packet flow for QoS handling by routers.	Does not require manual configuration or DHCP.
12.	Both routers and the sending host fragment packets.	Uses host address (AAAA) resource records in DNS to map host names to IPv6 addresses.
13.	Header includes a checksum.	Uses pointer (PTR) resource records in the IP6.ARPA DNS domain to map IPv6 addresses to host names.
14.	IPSec support is optional.	Must support a 1280-byte packet size (without fragmentation).

3.11. INTERNET COMMUNICATION TOOL

The internet communication takes place in following three ways and communication tools are selected on the bases as follows:

1. One-to-One Communication

This Type of communication is very common on internet and we can select following tool for it: Email, chat-type systems are the example.

2. Large and Small Group Communication

We can put following tool in this category:

- Social Networks
- Blogging
- Instant Messaging
- Chat
- Usenet
- Mail-based discussion list

3. Mass Communication

We can select following tool:

- Net Video
- Net Audio
- Mass media outlets

SUMMARY

- Now a days internet is backbone of any society. In this chapter we have gone through the basic working and infrastructure requirement of internet.
- Students can easily understand about Internet addressing, ISP, Modems, Internet Connectivity, and Modes of Internet Connection. Chapter discussed in detail about IPv4 and IPv6 addressing schemes of the Internet with their comparison.
- To connect to the net, we run a communication program to dial the number of an Internet service provider. When the remote system answers the phone, your modem talks to the other system and a connection is established.
- The Internet's growth has become explosive and it seems impossible to escape the bombardment of www.com's seen constantly on television, heard on radio, and seen in magazines. Because the Internet has become such a large part of our lives, a good understanding is needed to use this new tool most effectively.

SHORT QUESTION ANSWERS

Q. 1. What is ISP (Internet Service Providers)?

Ans. For networks connected to the global Internet, an organization obtains network number from the communication company that supplies internet connections. Such companies are called Internet Service Provider (ISPs).

Q. 2. Define Modems?

Ans. A modem is a device or program that enables a computer to transmit data over, for example, telephone or cable lines. Computer information is stored digitally, whereas information transmitted over telephone lines is transmitted in the form of analog waves. A modem converts between these two forms.

Q. 3. Explain different types of Internet Account?

Ans. There are two possible types of connections :

1. Shell Account
2. PPP (point to point protocol) account.

Q. 4. What is Telephone Line Internet Connectivity (Dial-up connection)?

Ans. The most basic type of Internet connection is called a dial-up connection. This connection is made through a modem (the communication mechanism in computers) that uses a telephone line to connect to the Internet. The modem must dial the telephone every time it wants to connect to the Internet, hence the name **dial-up connection.**

Q. 5. What is ISDN - Dial-Up Connection?

Ans. ISDN (Integrated Services Digital Network) connection is a high-speed dial-up and requires a special type of dedicated *digital* telephone line. This type of connection improves speed because the signals received are already in a digital format, but it is costly. For a while dial-up connections were the only way to access the Internet.

Q. 6. What is Broadband Connections?

Ans. Broadband is a high-speed Internet connection that makes surfing the Web more enjoyable. It also easily accommodates the video, audio, or complex graphics that are becoming commonplace on the Internet. Broadband is a type of data transmission in which a single medium or wire can carry several channels or communication paths at once.

Q. 7. What is Internet Connection Through Cable?

Ans. Internet over cable is a technology and service that uses the existing cable TV networks (made up of coaxial cable lines that bring television signals to TV) to bring Internet data to PC or TV at a very high speed. The television and the Internet transmission take place simultaneously on the same cable but at different frequencies. This allows the user to view TV and access Internet at the same time.

Q. 8. What is Internet Connectivity through Leased Line?

Ans. Leased lines are dedicated circuits provided by Basic Service Providers (BSPs), which provide permanent connectivity to the Internet. Leased lines provide the last mile access from the user premises to the ISP. They provide permanent connection as compared to the temporary connectivity through dialup access. The quality of the connection is far superior to what is normally available through dialup, thanks to digital signaling, less noise, fewer exchanges etc.

Q. 9. What is Internet Protocol version 4 (IPv4)?

Ans. IPv4 is the fourth revision in the development of the Internet Protocol (IP) and it is the first version of the protocol to be widely deployed. IPv4 is a connection-less protocol to be used on packet-switched Link Layer networks (e.g., Ethernet)".

Q. 10. What is IPv6?

Ans. The Internet Engineering Task Force (IETF) has designated IPv6 as the successor to version 4 for general use on the Internet. It significantly increases the size of the address space used to identify communication endpoints in the Internet, thereby allowing it to continue its tremendous growth rate. IPv6 is also known as IPng (IP Next Generation).

REVIEW QUESTIONS

1. What is ISP? How you will select an ISP ?
2. What is Domain Name System ?
3. How addressing on internet takes place ?
4. Compare various addressing schemes for internet ?
5. Discuss various types of Internet accounts and their drawback ?
6. Discuss various modes of internet connectivity ?
7. How can be select internet communication tool ?
8. Write the short notes on following :

   (a) DNS (b) IPv4
   (c) IPv6 (d) Broadband
   (e) Leased Line (f) Modem
   (g) URL

❑❑❑

WORLD WIDE WEB

SECTION-B

- **World Wide Web**
- **HTML and XML**
- **Web Site Designing**

SYLLABUS

Introduction, Miscellaneous Web Browser Details, Searching the WWW: Directories Search Engines and Meta Search Engines, Search Fundamentals, Search Strategies, Working of the Search Engines, Telnet and FTP. Introduction to Browser, Coast-to-Coast Surfing, Hypertext Markup Language, Web Page Installation, Web Page Setup, Basics of HTML and Formatting and Hyperlink Creation. Using FrontPage Express, Plug-ins.

4

CHAPTER

World Wide Web

4.1. INTRODUCTION [MDU Dec. 2006]

INSIDE THIS CHAPTER

Servers store hypertext markup language (HTML) files and respond to requests. Through the use of a browser, PC users can find and view server-based documents. Browsers like Netscape ensure easy access to server-based documents and display multimedia data. Future browser will include editor-like abilities, application linking, audio and video integration; data based front end, and ways to combine information with transactions.

Reason for the amazing success of the Web include :

1. Ease of navigation and use.
2. Ease of publishing content.
3. New distribution models.
4. Enabling a network-centric computing paradigm.
5. Enabling new intra-business applications.

Definition

World Wide Web is broad term, which is the collection of distributed documents referred to as "pages" located on computers (or servers) all over the world. For example : http://kiet.edu.

4.1.1. Web Architecture

To use the Web, in addition to an Internet connection, a user need a special piece of software called a Web browser (such as Netscape Navigator). The browser act as a graphical interface between the user and the Internet – it sends the necessary command to request data from other computers and then format them for the user's screen. Documents that are formatted using hypertext markup language (HTML) contain taglines that inform the browser how to format them.

The web is based on a three-part architecture :

1. **HTML** : Hypertext markup language, the format for web pages, provide both formatting and hyperlinking.
2. **HTTP** : Hypertext transfer protocol, the protocol for communications between Web servers and browsers.
3. **CGI** : The common gateway interface, the interface for invoking programs from web servers.

Figure 4.1 shows the component of web architecture, including the web clients, web servers, HTTP protocol language for Web client/server communication and applications. Within the confines of this architecture, hundreds of entrepreneurs have produced tidal wave of software programs and tools.

The web architecture forms the basis for electronic commerce applications that involve building software in which the functions are distributed among :

1. Application servers (where application reside),
2. Data servers (where most of the data resides), and
3. A Group of client computers that are usually networked PCs (where the information user work).

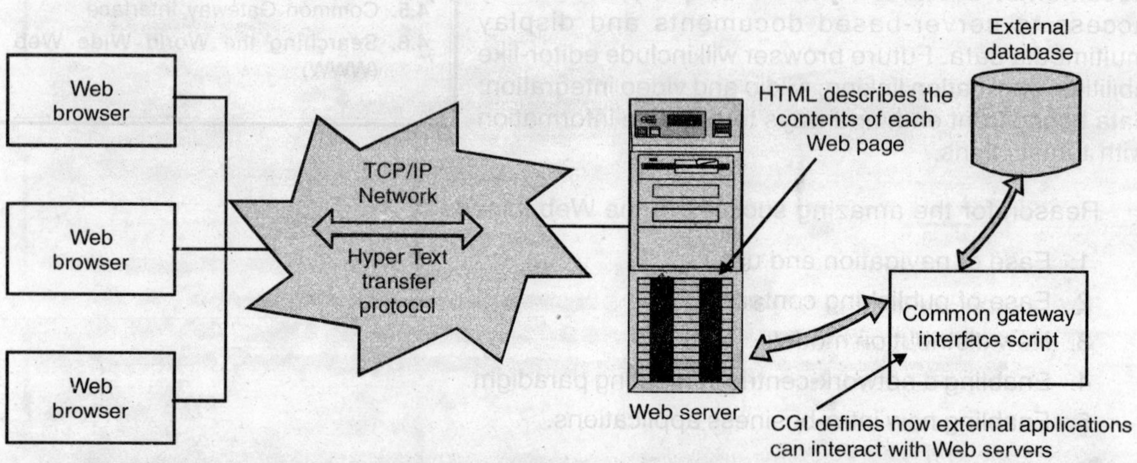

FIGURE 4.1. The Web Architecture.

4.2. TCP/IP PROTOCOLS

There are 4 protocol layers that data goes through before it gets sent off to another computer. If the message to be sent is long, it will be broken into smaller chunks called **packets**.

TIPS

We know that rules are very important to make the communication possible and TCP/IP are the protocols which make the communication possible over internet.

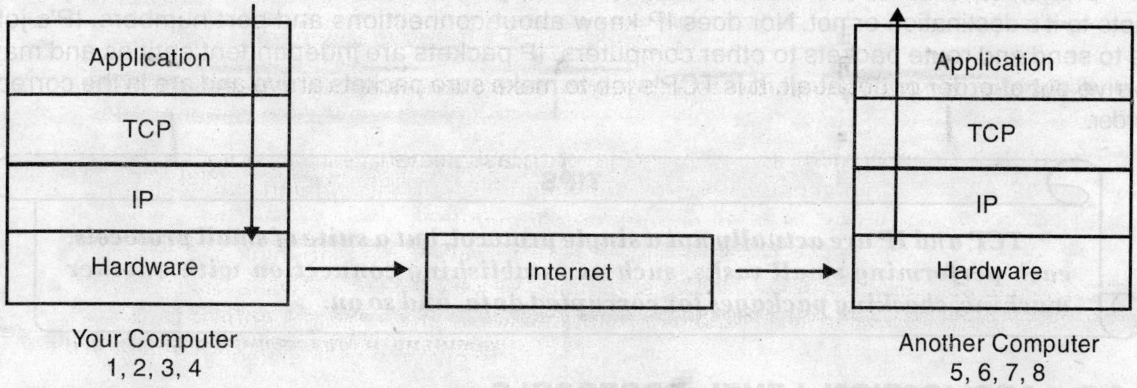

FIGURE 4.2. TCP/IP Protocol.

1. **Application Protocols Layer** : Protocols specific to applications such as WWW, e-mail, FTP, etc.
2. **Transmission Control Protocol Layer** : TCP attaches a **port number** of the destination service to packets to be sent out, and directs arriving packets to a specific application on a computer using a port number. Every internet service communicates at a specific port. Ports are integer numbers which allow a computer to distinguish between several internet communications happening at the same time, but facilitates by different internet applications. For instance, since FTP and HTTP use different ports, a computer can be downloading a file and displaying web pages at the same time. Most common ports are :
   - FTP : 20/21
   - Telnet : 23
   - HTTP : 80

   TCP also handles assembly and disassembly of long messages into packets (smaller chunks of data). Each packet will be "wrapped" separately by all the protocols that the packet goes through. It will also get a sequence number, so that the packets will be reassembled at the destination (packets may arrive out of order). The last packet is marked to indicate that the message is complete.
3. **Internet Protocol Layer** : IP converts the domain name into IP address, attaches the IP address of the sending machine, directs packets to a specific computer using an IP address.
4. **Hardware Layer** : Converts binary packet data to network signals and back (For example, ethernet network card, modem for phone lines, etc.). That's how the data actually gets transmitted.

TCP is a Connection-Oriented, Reliable, Byte Stream Service : Connection-oriented means that two applications using TCP must first establish a connection before exchanging data. TCP is reliable because for each packet received, an acknowledgement is sent to the sender to confirm the delivery. TCP also includes a checksum in it's header for error-checking the received data. It sends data in byte streams, not character streams, meaning that any binary data (pictures, Excel files, etc.) may be sent using TCP.

Unlike TCP, **IP is an unreliable, connectionless protocol**. IP doesn't care whether a packet gets to it's destination or not. Nor does IP know about connections and port numbers. IP's job is to send and route packets to other computers. IP packets are independent entities and may arrive out of order or not at all. It is TCP's job to make sure packets arrive and are in the correct order.

TIPS

TCP and IP are actually not a single protocol, but a suite of small protocols, each performing small tasks, such as establishing connection with another machine, checking packages for corrupted data, and so on.

4.3. APPLICATION LEVEL PROTOCOLS

Application level protocols, s.a. HTTP or FTP, build on top of TCP and IP.

4.3.1. HTTP

Definition

HTTP is a connectionless text based protocol to make the communication possible over web.

Clients (web browsers) send requests to web servers for web elements such as web pages and images. After the request is serviced by a server, the connection between client and server across the Internet is disconnected.

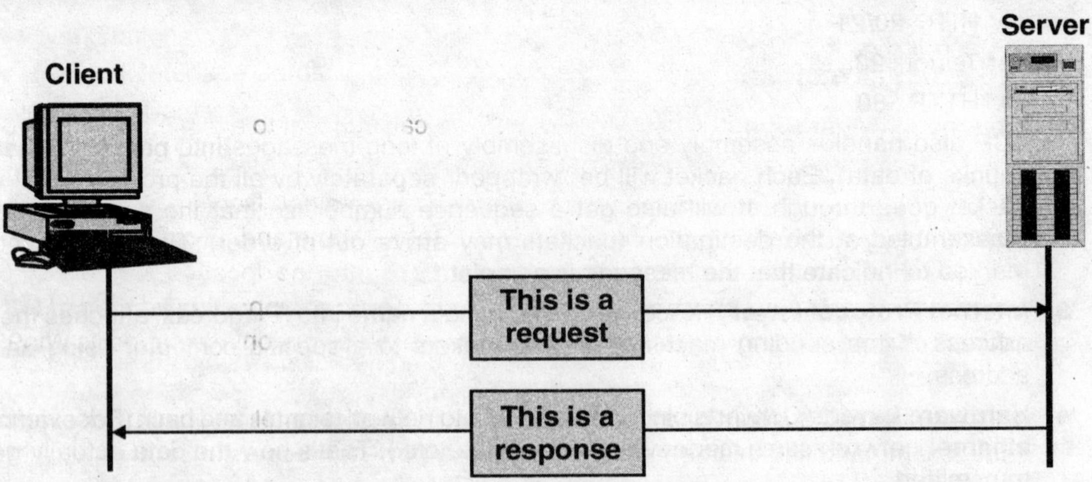

FIGURE 4.3. HTTP Request and Response.

A new connection must be made for each request. This means that HTTP is not connection-oriented. On the practical side, it means that after a requested HTML file gets downloaded, your computer is disconnected from the server, then your browser discovers that it needs a .gif image for the page, makes another request to the server, disconnects again and so on.

Each HTTP interaction works in following ways :

1. The browser established a connection with the server.
2. The browser send a request to server.
3. The server sends response to the client (browser).
4. Both browser and server disconnect and transaction ends.

4.3.2. Telnet Protocol [MDU Dec. 2006]

Definition

Telnet is method of connecting to a remote computer over the Internet that let us use programs and data just as if we were using the computer locally. Telnet is a text-only protocol.

At one time it was one of the most common ways to connect to other sites. Now with the growth of the web, it is hardly used. There are still a few sites and resources that can only be reached through telnet.

Use of Telnet

- If we are a gamer, there are a number of text based games still available through telnet. See Yahoo's page on MUDs (Multi-User Dungeons).
- Outside of games, unless we already know of a particular use, such as connecting to a particular government database, we probably have no use for it.
- For years, many library catalogs were only reachable through telnet. Many early community resources, called free-nets, were reachable through telnet. You would have trouble finding any of these now.

Connect to a Telnet Address

If our Internet provider supports telnet (not all do), we can connect to a telnet address in one of two ways :

- Recent versions of Windows include a command line telnet. Begin at Start/Run and type **command**. On the next screen, type your telnet command. From a SLIP or PPP or ethernet connected account: type a telnet URL into the location window of our browser. To do this, our browser must be configured with a supporting version of telnet software. If we are using Windows95, make sure we get a version that supports Win95's 32-bit stack.
- From a unix shell account or other account that lets you type telnet commands, we can issue a command like this: **telnet some.address port-number**

4.3.3. File Transfer Protocol [MDU Dec. 2006, May 2008]

Definition

FTP is means of communication over the network specifically designed for moving files between computers on the network. Primary use to most of us is transferring files to/from our desktop/laptop personal computers and the unix workstations, but it can also be used to communicate with machines at companies like Netscape to download software, or any PCs or other machines running FTP servers.

FTP servers listen on port 21. Data connections are initiated by the server from its port 20, to a port on the client identified in a PORT command. Following are recommended Programs for FTP :

Mac:	Fetch	Nice graphical interface that supports drag-and-drop
PC:	WS-FTP	Pretty good graphical interface once you get used to it.
	built-in	Windows 95 built-in support for ftp, command-line oriented.
Unix:	ftp	Standard Unix command-line ftp program, virtually identical to the PC built-in.
	ncftp	Slightly more feature-packed version, also command line oriented.

Using FTP to Download (Transfer from Another Machine to Yours)

1. **Connect to the Server (and Login) :** Servers with restricted access requires a username and a password. Other servers (such as ftp.netscape.com) support "anonymous login" where the standard is to use the username "anonymous" and our full email address as your password.
2. **Find the File(s)/Folder(s) We Want to Download :** Navigating the folder hierarchy is just like it is on Macs, PCs, etc. Double-click a folder to open it. (Command-line FTP programs are more like navigating in dos/unix; use the "cd" command).
3. **Choose the Appropriate Download Mode :** There are two transfer modes in FTP. The first is called either "ascii" or "text", and the other is called "binary" or "image". The reason for the two modes is that different machines represent End-Of-Line characters in different ways. "text" mode automatically corrects for these differences so that the text we download appears correct on our machine, so for text files we should use text mode (otherwise, we may find, for example, that all our text appears on one line after it is downloaded). But if we use text mode on a non-text file, it will corrupt the file. It will try to interpret non-text data as text and therefore change the file. And for something non-text like a graphic or a microsoft word document (which contains formatting), you will find very unpleasant results if you use text mode! In short, if it isn't text, use binary mode (which leaves the file untouched).
4. Begin the transfer and get the file (make sure we know where the file is being put on our local machine).

Using FTP to Upload (Transfer from Your Machine to Another)

1. **Connect to the Server (and Login) :** Servers with restricted access require a username and a password. Other servers (such as ftp.netscape.com) support "anonymous login" where the standard is to use the username "anonymous" and our full email address as our password.
2. If we are using Fetch, choose to "Put" files/folders, and we will be prompted to specify the files.
3. Find the file(s)/folder(s) we want to upload on your computer. Navigating the folder hierarchy is just like it is on Macs, PCs, etc. double-click a folder to open it.
4. **Choose the Appropriate Upload Mode!** See the notes on download for details. Note that Fetch chooses different names for the file types when uploading. "Raw Data" is the same as binary/image. "Text" is text.
5. Begin the transfer.

4.3.4. The SMTP Protocol

Definition

SMTP (Simple Mail Transfer Protocol) is a TCP/IP protocol used in sending and receiving e-mail.

However, since it's limited in its ability to queue messages at the receiving end, it's usually used with one of two other protocols, POP3 or Internet Message Access Protocol, that let the user save messages in a server mailbox and download them periodically from the server. In other words, users typically use a program that uses SMTP for sending e-mail and either POP3 or IMAP for receiving messages that have been received for them at their local server.

SMTP usually is implemented to operate over Transmission Control Protocol port 25. Simple Mail Transfer Protocol (SMTP), is Internet's standard host-to-host mail transport protocol and traditionally operates over TCP, port 25. In other words, a UNIX user can type telnet hostname 25 and connect with an SMTP server, if one is present.

If mail delivery fails, sendmail (the most important SMTP implementation) will queue mail messages and retry delivery later. However, a backoff algorithm is used, and no mechanism exists to poll all Internet hosts for mail, nor does SMTP provide any mailbox facility, or any special features beyond mail transport. For these reasons, SMTP isn't a good choice for hosts situated behind highly unpredictable lines (like modems). A better-connected host can be designated as a DNS mail exchanger, then arrange for a relay scheme. Currently, there two main configurations that can be used. One is to configure POP mailboxes and a POP server on the exchange host, and let all users use POP-enabled mail clients. The other possibility is to arrange for a periodic SMTP mail transfer from the exchange host to another, local SMTP exchange host which has been queuing all the outbound mail. Of course, since this solution does not allow full-time Internet access, it is not too preferred.

4.3.5. Secure Socket Layer Protocol

Definition

SSL is a layered protocol. It is located above a reliable transport layer, such as TCP, and below an application layer. However, SSL is not TCP\IP dependant and can run under other application protocols. It takes applications' data to be sent, fragments it into manageable blocks, compresses it (optionally), applies a Message Authentication Code (MAC) to the data, encrypts it and then transmits the result.

When data is received, the reverse process occurs: the data is decrypted, verified, decompressed, reassembled and then delivered to the application layer.

The SSL protocol consists of two major layers :

(1) SSL Handshake protocol and

(2) SSL Record protocol.

In addition, there are two other elements :

(1) SSL Change Cipher Spec Protocol, and

(2) SSL Alert Protocol.

The Handshake and the latter two components are responsible for the management of SSL communication setup and security parameters. The SSL Record Protocol is the lowest layer of the SSL protocol stack and all the other parts, including application data, operate on top of it. The Figure 4.4 shows the SSL protocol stack.

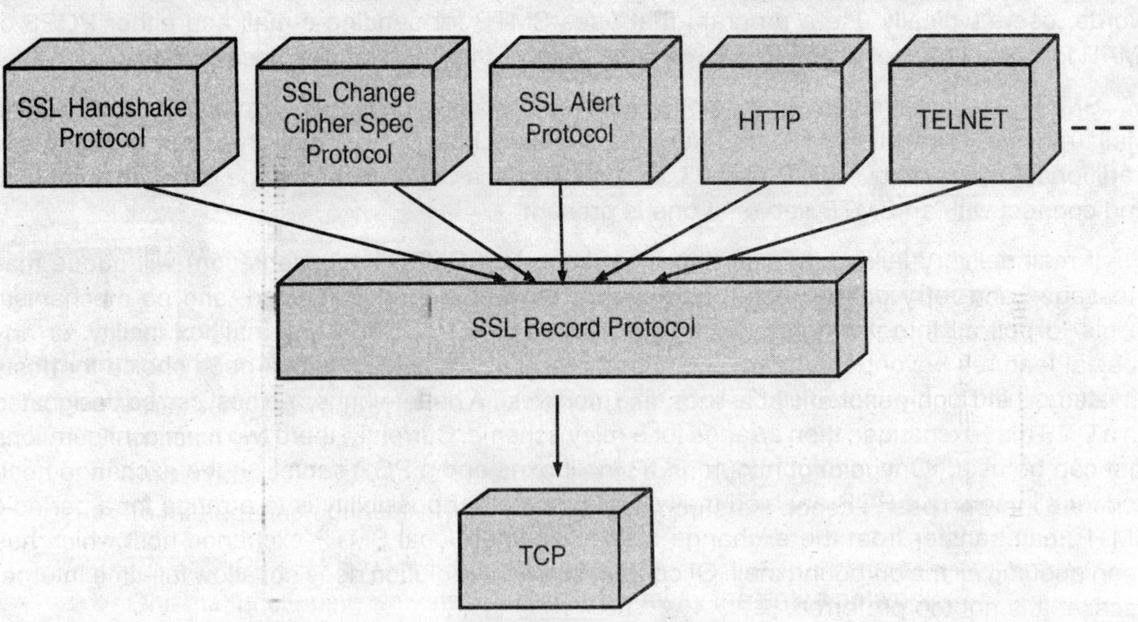

FIGURE 4.4. The SSL Protocol Slack.

4.4. WEB BROWSER [MDU May 2008]

Definition ***Web user access information from web servers, through a client program called browser.***

Broadly speaking, a browser is responsible for the following tasks :

1. The first task is to accept a URL and retrieve the document specified by the URL.
2. The second task of browser is to interpret a HTML, document code, format it accordingly and finally render the document on the screen.
3. In this process, it may have to manage the rendering of various image formats, multimedia information, and links to other documents as well.

TIPS

The most Famous Example of Browsers are Microsoft Internet Explorer and Netscape Navigator.

Information regarding downloading the Netscape browser can be found at the site http://www.netscape.com. It can be download from the site or any of the listed mirror sites, and closest to your location. The internet explorer of Microsoft is popular browser, based on the Mossaic source code.

The overview of the architecture of the web browser is shown in following Figure 4.5 :

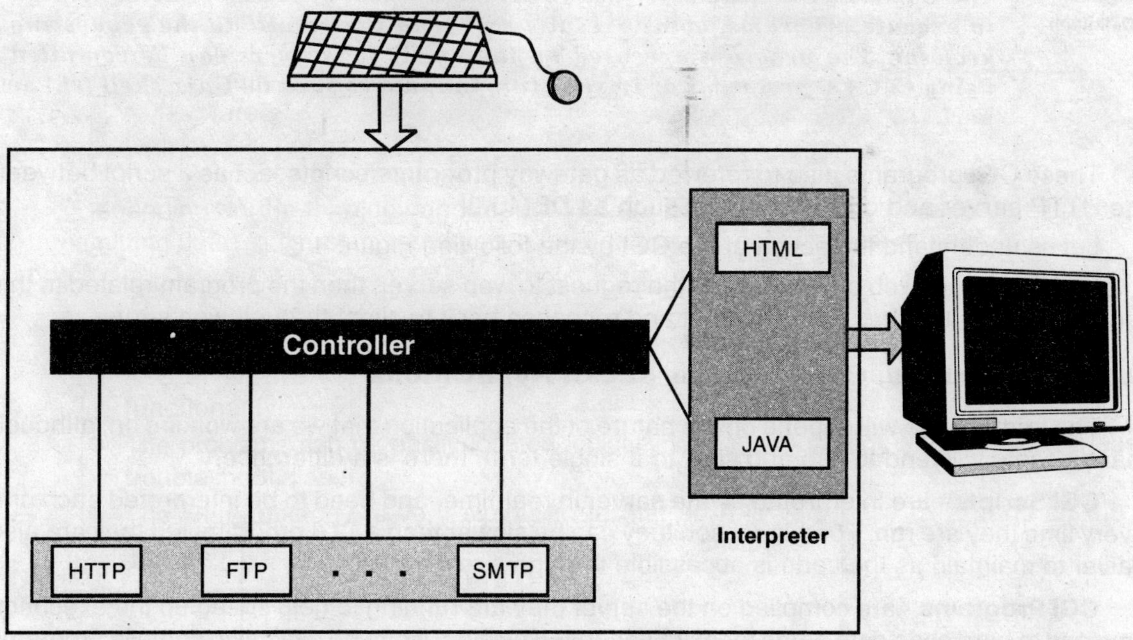

FIGURE 4.5. **Browser Architecture.**

4.5. COMMON GATEWAY INTERFACE

The server manages HTML documents and delivers them to the browser, on request. These pages delivered by the server are usually HTML documents that have been prepared and placed on the server. For example, a list of items available in the company can be coded in a HTML document and placed on the server along with the inventory position. The change in the item inventory database need to be reflected on the product offering web page, served to the browser. Static HTML pages will require modification of the document every time the inventory database undergoes a change. A better solution would be to have a mechanism that generates the list of item document from the inventory database for each request of the document by any client/browser.

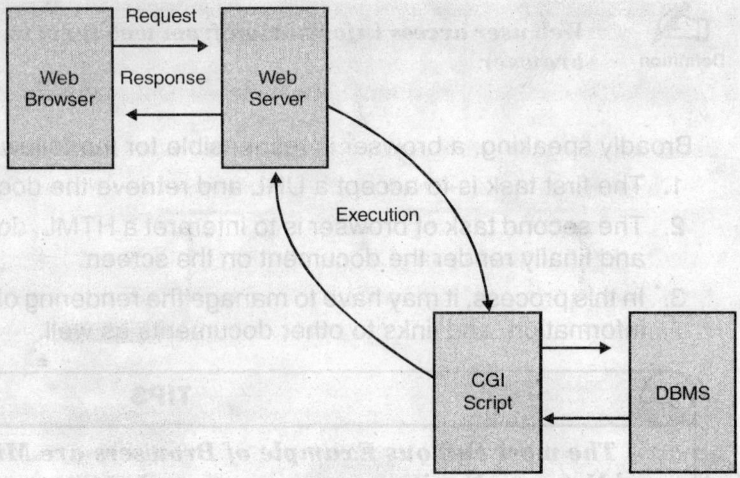

FIGURE 4.6. Interaction Between Browser and CGI Program.

Definition

The Common Gateway Interface (CGI) mechanism of HTTP enables servers to execute programs, obtain results and send the results to the requesting browser. The program executed by the by HTTP servers can be compiled using C, C++ programs or in scripting languages such as Unix shell (sh) or perl.

These CGI programs also to referred as gateway programs/scripts, act as a script between the HTTP server and other programs such as DBMS."

Let us understand the action of the CGI by the following Figure 4.6.

In above figure web browser send the request to web server, then the program related to this request is executed through CGI script and response back to client through web server.

4.5.1. CGI Scripts, CGI Programs or CGI Applications

The term we use will depend on the nature of the application that we are working on, although many people will tend to use and stick to a single term, there is a difference.

CGI Scripts - are interpreted by the server in real-time, and need to be interpreted each and every time they are run. For this reason they can be slower than a CGI program, but they are also easier to maintain as the code is accessible through a text editor.

CGI Programs - are compiled on the server they are running to help speed up the execution time and to hard code data paths for extra security. However, it can be difficult to compile programs on systems where access is limited, and the program will need to be recompiled for each platform we want to use it on.

CGI Applications - typically a CGI Application is a collection of CGI scripts or programs that work together to create a web-based interface for a program or database, although a single stand alone program or script can be described as an application if it does more than one type of server interaction.

4.6. SEARCHING THE WORLD WIDE WEB (WWW)

Searching the World Wide Web can be both beneficial and frustrating. We may find vast amounts of information, or we may not find the kinds of information we're looking for. Searching online will provide you with a wealth of information, but not all of it will be useful or of the highest quality.

With billions of web pages online, you could spend a lifetime surfing the Web, following links from one page to another. Amusing perhaps, but not very efficient if you're seeking some specific information. So where do you start ?

FIGURE 4.7. Web Searching.

Definition

Searching the Internet requires part skill, part luck and a little bit of art.

The World Wide Web is a superb resource, but it doesn't contain all the information that you can find at a library or through library online resources. Don't expect to limit your search to what is on the Internet, and don't expect search engines to find everything that is on the Web.

Studies of search engine usage show that search engines are increasing exponentially in their indexing of new websites and information. Indexing is the web term for finding and including new web pages and other media in search results. For example, in 1994, Google indexed approximately 20 million pages. As of 2004, that number is up to 8 billion! However, search engines still only index a fraction of what is available on the Internet and not all of it is up to date. Search engines may only "crawl" sites (or revisit them for purposes of indexing) every month or so; information that has been updated since that time will be invisible to the search engines. After you try several search engines, you will see that you get different results from different sites. Also, remember that some information appears and then disappears from Websites. Finally, search engines don't always search the entire page; if a page is larger than 100 to 500 k, many search engines will only index the first 100 to 500k of the page. So there could be valuable information that is being overlooked by a search engine even in pages that are indexed.

Not all of the information located on the Internet is able to be found via search engines. Researchers Chris Sherman and Gary Price call this information the "invisible web" (another name that is frequently used is the "deep web"). Invisible web information includes certain file formats, information contained in databases, and other omitted pages from search engines.

So, using search engines is not the only way to find material on the web, but they are one tool you can use. Knowing a few search strategies and hints, as you use these engines, can make the search more profitable.

4.6.1. Web Directory and Search Engines [MDU Dec. 2007, May 2008]

Definition

Search directories are hierarchical databases with references to websites. The websites that are included are handpicked by living human beings and classified according to the rules of that particular search service.

There are two ways for sites to be included in a Web directory's listings; either the site owner can submit the site to the Web directory, or the directory's editor(s) will eventually come across that site.

The example of web directories are as follows:

- Yahoo has one of the best Web directories on the Web, along with its own search engine listings and many other search services.
- Look Smart is hard to classify as just a Web directory; especially with its recent foray into vertical search channels.

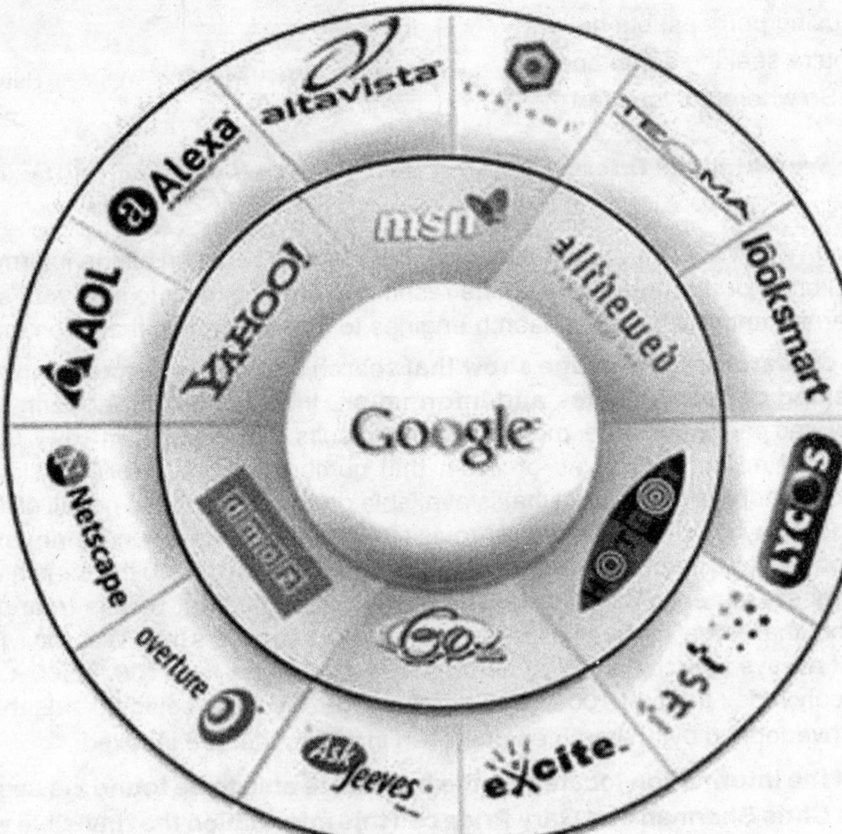

FIGURE 4.8. Search Engines.

- The Open Directory has been around since 1999, and is a human-edited directory. Also known as DMOZ (Directory Mozilla), the Open Directory Project suppose to be the "largest on the Web", constructed and maintained by a "vast, global community of volunteer editors."

Definition

Search Engine is a software program that searches for sites based on the words that you designate as search terms. Search engines look through their own databases of information in order to find what it is that you are looking for. Examples of search engines are Google, AltaVista, Atomz, FreeFind and SpiderLine.

Search engines and directories are not the same thing; although the term "search engine" often is used interchangeably. Search engines automatically create web site listings by using spiders that "crawl" web pages, index their information, and optimally follows that site's links to other pages. Spiders return to already-crawled sites on a pretty regular basis in order to check for updates or changes, and everything that these spiders find goes into the search engine database.

4.6.2. Meta Search Engines [MDU May 2005, 2007]

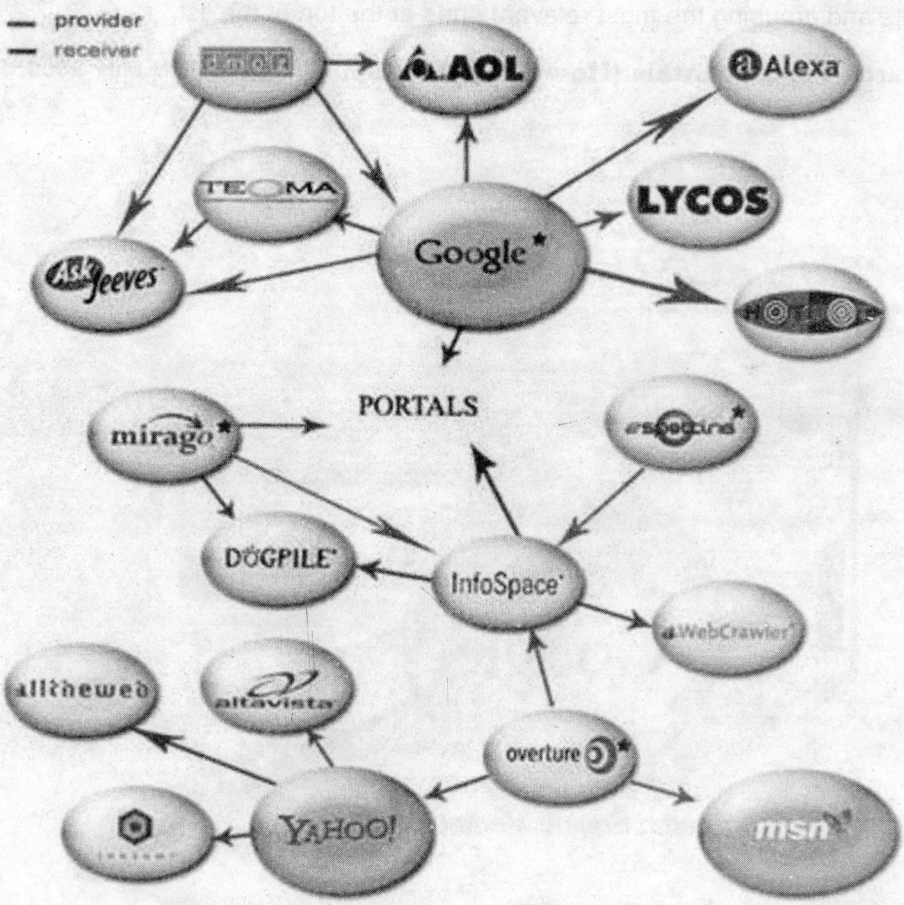

FIGURE 4.9. Meta Search Engine Technique.

Definition

A meta-search engine is a search tool that sends user requests to several other search engines and/or databases and aggregates the results into a single list or displays them according to their source.

Meta search engines enable users to enter search criteria once and access several search engines simultaneously. Meta search engines operate on the premise that the Web is too large for any one search engine to index it all and that more comprehensive search results can be obtained by combining the results from several search engines. This also may save the user from having to use multiple search engines separately.

Meta search engines create a virtual database. They do not compile a physical database or catalogue of the web. Instead, they take a user's request, pass it to several other heterogeneous databases and then compile the results in a homogeneous manner based on a specific algorithm.

No two Meta search engines are alike. Some search only the most popular search engines while others also search lesser-known engines, newsgroups, and other databases. They also differ in how the results are presented and the quantity of engines that are used. Some will list results according to search engine or database. Others return results according to relevance, often concealing which search engine returned which results. This benefits the user by eliminating duplicate hits and grouping the most relevant ones at the top of the list.

4.6.3. Search Fundamentals (How They Works) [MDU May 2006, Dec. 2007]

FIGURE 4.10. Searching on Search Engines.

The internet contains a vast collection of information, which is spread out in every part of the world on remote web servers. The probiem in locating the correct information on the internet led to the creation of search technology, known as the **internet search engine**. A search engine can provide links to relevant information based on your requirement or query. In order to understand the terminology and techniques to position our website pages for higher ranking in search engines, the knowledge of the basic functioning of a search engine is essential.

Functions of Internet Search Engines

Each search engine does the same functions of collecting, organizing, indexing and serving results in its own unique ways, thus employing various algorithms and techniques, which are their trade secrets. In short, the functions of a search engine can be categorized into the following :

1. Crawling the internet for web content.
2. Indexing the web content.
3. Storing the website contents.
4. Search algorithms and results.

4.6.3.1. Crawling and Spidering the Web

Definition

Crawling is the method of following links on the web to different websites, and gathering the contents of these websites for storage in the search engines databases.

*The **crawler** (also known as a web robot or a web spider) is a software program that can download web content (web pages, images, documents and other files), and then follow hyper-links within these web contents to download the linked contents. The linked contents can be on the same site or on a different website.*

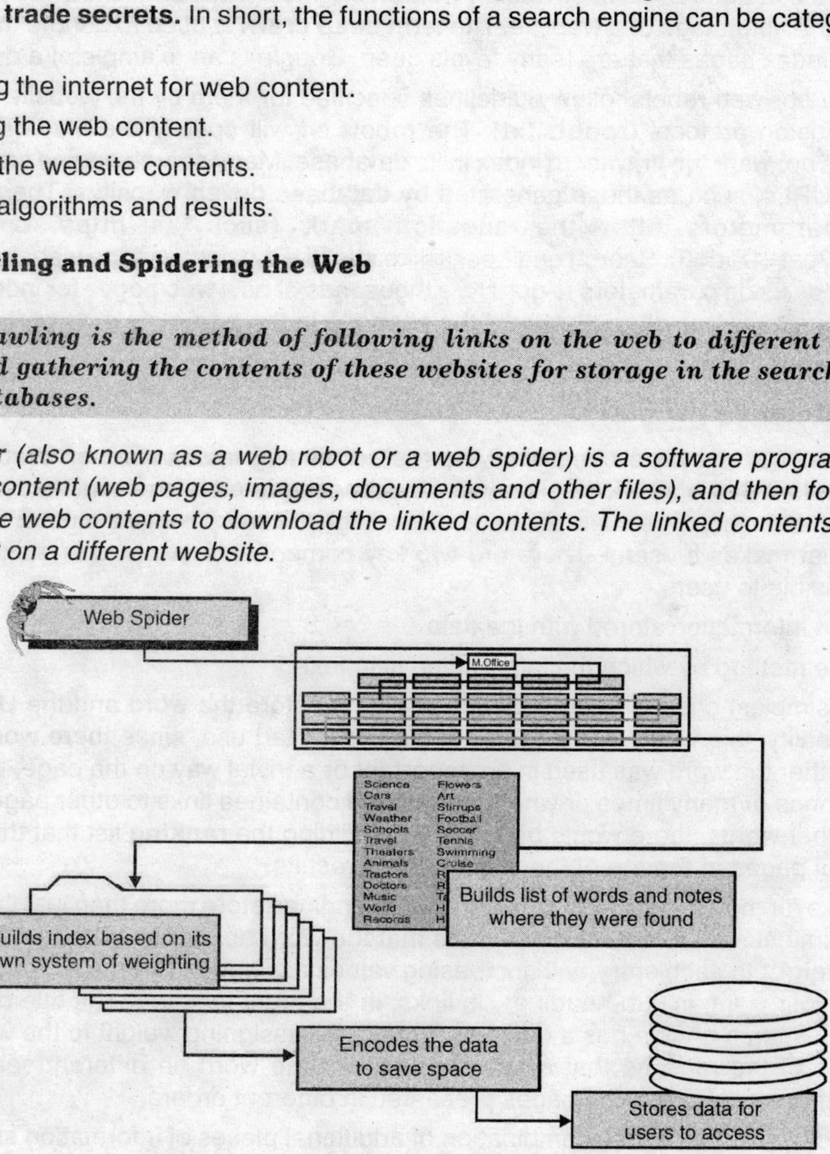

FIGURE 4.11. Crawling and Spidering the Web.

The crawling continues until it finds a logical stop, such as a dead end with no external links or reaching the set number of levels inside the website's link structure. If a website is not linked from other websites on the internet, the crawler will be unable to locate it. Therefore, if the website is new, and has no links from other sites, that website has to be submitted to each of the search engines for crawling.

The efficiency of the crawler makes it crawl multiple websites at the same time, so as to collect billions of website contents as frequently as it can. News and media sites are crawled more frequently (every hour or so) by advanced search engines like Google, in order to deliver updated news and content in their search results. The crawler also does not flood a single website with a high volume of requests at the same time, but spreads the crawling over a period of time so that the web site does not crash. Usually search engines crawl only a few (three or four) levels deep from the homepage of a website. The term **deep crawl** is used to denote that the crawler or spider can index pages that are many levels deep. Google is an example of a deep crawler.

Crawlers or web robots follow guidelines specified for them by the website owner using the robots exclusion protocol (**robots.txt**). The robots.txt will specify the files or folders that the owner does not want the crawler to index in its database. Many search engine crawlers do not like unfriendly URLs, such as those generated by database driven websites. These website URLs contain parameters after the question mark (such as http://somedomain.com/article.php?cat=1&id=3). Search engines dislike such URLs because the website can overwhelm the crawler by using parameters to generate thousands of new web pages for indexing with similar content. Thus, crawlers often disregard the changes in the parameters as part of a new URL to spider.

4.6.3.2. Building the Index

Once the spiders have completed the task of finding information on Web pages (and we should note that this is a task that is never actually completed — the constantly changing nature of the Web means that the spiders are always crawling), the search engine must store the information in a way that makes it useful. There are two key components involved in making the gathered data accessible to users:

- The information stored with the data
- The method by which the information is indexed

In the simplest case, a search engine could just store the word and the URL where it was found. In reality, this would make for an engine of limited use, since there would be no way of telling whether the word was used in an important or a trivial way on the page, whether the word was used once or many times or whether the page contained links to other pages containing the word. In other words, there would be no way of building the **ranking** list that tries to present the most useful pages at the top of the list of search results.

To make for more useful results, most search engines store more than just the word and URL. An engine might store the number of times that the word appears on a page. The engine might assign a **weight** to each entry, with increasing values assigned to words as they appear near the top of the document, in sub-headings, in links, in the meta tags or in the title of the page. Each commercial search engine has a different formula for assigning weight to the words in its index. This is one of the reasons that a search for the same word on different search engines will produce different lists, with the pages presented in different orders.

Regardless of the precise combination of additional pieces of information stored by a search engine, the data will be **encoded** to save storage space. For example, the original Google paper describes using 2 bytes, of 8 bits each, to store information on weighting — whether the word was capitalized, its font size, position, and other information to help in ranking the hit. Each factor might take up 2 or 3 bits within the 2-byte grouping (8 bits = 1 byte). As a result, a great deal of information can be stored in a very compact form. After the information is compacted, it's ready for indexing.

An index has a single purpose: It allows information to be found as quickly as possible. There are quite a few ways for an index to be built, but one of the most effective ways is to build a **hash table**. In **hashing**, a formula is applied to attach a numerical value to each word. The formula is designed to evenly distribute the entries across a predetermined number of divisions. This numerical distribution is different from the distribution of words across the alphabet, and that is the key to a hash table's effectiveness.

4.6.3.3. Component of Search Engines

There are four main components to a search engine which will affect the results of your search.

(1) The Webcrawler

This is the part of the search engine which combines the pages on the internet and gathers the information for the search engine. Variable features which can affect your search results include :

- **Included Pages :** Most search engines will find information by beginning at one page and then following all of the links on that page. It will then follow all of the links on these new pages, and so on. Therefore, if a page is not linked to from another page, it may never be found by a search engine. Authors can include unlinked pages in a search engine by submitting them to each specific search engine.
- **Excluded Pages :** Some web administrators may choose to exclude their pages from search engines because they are internal pages or intranets. Many web pages are also excluded because their content is dynamically generated from a database and a search engine cannot find it.
- **Documents Types :** Different search engines will search different document types. All will search HTML documents, but some will also search PDF, PowerPoint, Word, Excel and more.
- **Frequency of Crawling :** An important part of a web crawler is how frequently it retrieves information from pages. Some sites it will visit more often than others.

(2) The Database

The search engine's database is what you are actually searching. All of the information that a web crawler retrieves is stored in a database. Every time you use a search engine, it is this database you are searching, not the live internet. Variable features which can affect your search results include :

- **Size of the Database :** Some search engines will have extremely large databases (~5 billion pages) for you to search, while others will have comparatively small ones (~150 million pages). A small database is not necessarily worse, however, since it may offer more focused and higher quality results then a larger database.
- **Freshness of the Database :** The freshness of the database is a direct result of how frequently the web crawler retrieves new information. If the information in the database is fairly old, then your search results will suffer.

(3) The Database

Each search engine interprets the terms you enter into the search box in different ways. Variable features which can affect your search results include :

- **Operators :** Most search engines allow you to use operators such as AND, OR and NOT in order to create complex search statements. The terms may need to be entered in upper case. Most also use the plus (+) sign to signal that a term must be included in the search results and a minus (–) sign to signal that a term must be excluded from the results.
- **Phrase Searching :** Search engines will generally search for words as phrases when marks are placed around the phrase. Some search engines may use a drop down menu which offers the option of searching the terms as a phrase.
- **Truncation :** Some search engines will automatically truncate the terms you enter. This means that the search engine will not only search for the term exactly as you spelled it, but will also search on that term with alternate endings and as a plural. Some search engines will only search for variable endings on certain common words.

(4) The Ranking Algorithm

How a search engine ranks the results of your search is possibly the most important component of a search engine. Most searches will retrieve thousands of results. Since you probably will only look through the first 1-2 pages of results, you need the most relevant results to appear first. Variable features which can affect your search results include :

- **Location and Frequency :** All search engines look at the location and frequency of words in a page. If a term appears near the top of a web page, such as in the title or in the first few paragraphs of text, it is assumed that the page is more relevant than if the term is used at the bottom of the page. Pages where the words appear more frequently in relation to the other words on the page also qualifies the page as being more relevant than other web pages.
- **Link Analysis :** This feature analyzes how pages link to each other and then uses this information to determine the "importance" of each page. If a page is linked to from a large number of other pages, then it is ranked more highly.
- **Clickthrough Measurement :** Some search engines also use clickthrough analysis. This means that a search engine might watch what results someone selects from a particular search, then eventually drop high-ranking pages that aren't attracting clicks, while promoting lower-ranking pages that do pull in visitors.

4.6.4. Search Strategies

A search engine is a device that sends out inquiries to sites on the web and catalogs any web site it encounters, without evaluating it. Methods of inquiry differ from search engine to search engine, so the results reported by each one will also differ. Search engines maintain an incredibly large number of sites in their archives, so you must limit your search terms in order to avoid becoming overwhelmed by an unmanageable number of responses.

Search engines are good for finding sources for well-defined topics. Typing in a general term such as "education" or "Shakespeare" will bring back far too many results, but by narrowing your topic, you can get the kind (and amount) of information that you need.

For example :

- Go to Google (a search engine)
- Type in a general term ("education")

- Add modifiers to further define and narrow your topic ("rural education Indiana")
- Be as specific as you can ("rural education Indiana elementary school")
- Submit your search.

Adjust your search based upon the number of responses you receive (if you get too few responses, submit a more general search; if you get too many, add more modifiers).

1. Learn how the Search Engine Works

Each search engine has different advantages. Google is one of the largest search engines, followed closely by MSN and Yahoo. This means that these three search engines will search a larger portion of the Internet than other search engines. Lycos allows you to search by region, language, and date. Altavista has searches for images, audio, video, and news. Ask Jeeves allows you to phrase your search terms in the form of a question. It is wise to search through multiple search engines to find the most available information.

2. Select Your Terms Carefully

Using inexact terms or terms that are too general will cause you problems. If your terms are too broad or general, the search engine may not process them. Search engines are programmed with various lists of words the designers determined to be so general that a search would turn up hundreds of thousands of references.

If your early searches turn up too many references, try searching some relevant ones to find more specific or exact terms. You can start combining these specific terms with NOT (see the section on Boolean operators below) when you see which terms come up in references that are not relevant to your topic. In other words, keep refining your search as you learn more about the terms.

Most search engines now have "Advanced Search" features. These features allow you to use Boolean operators (below) as well as specify other details like date, language, or file type.

3. Know Boolean Operators

Searching through an index involves a user building a **query** and submitting it through the search engine. The query can be quite simple, a single word at minimum. Building a more complex query requires the use of Boolean operators that allow you to refine and extend the terms of the search.

The Boolean operators most often seen are:

- **AND :** All the terms joined by "AND" must appear in the pages or documents. Some search engines substitute the operator "+" for the word AND.
- **OR :** At least one of the terms joined by "OR" must appear in the pages or documents.
- **NOT :** The term or terms following "NOT" must not appear in the pages or documents. Some search engines substitute the operator "-" for the word NOT.
- **FOLLOWED BY :** One of the terms must be directly followed by the other.
- **NEAR :** One of the terms must be within a specified number of words of the other.
- **Quotation Marks :** The words between the quotation marks are treated as a phrase, and that phrase must be found within the document or file.

4.6.5. Future Search

The searches defined by Boolean operators are **literal** searches — the engine looks for the words or phrases exactly as they are entered. This can be a problem when the entered words have multiple meanings. "Bed," for example, can be a place to sleep, a place where flowers are planted, the storage space of a truck or a place where fish lay their eggs. If you're interested in only one of these meanings, you might not want to see pages featuring all of the others. You can build a literal search that tries to eliminate unwanted meanings, but it's nice if the search engine itself can help out.

One of the areas of search engine research is **concept-based** searching. Some of this research involves using statistical analysis on pages containing the words or phrases you search for, in order to find other pages you might be interested in. Obviously, the information stored about each page is greater for a concept-based search engine, and far more processing is required for each search. Still, many groups are working to improve both results and performance of this type of search engine. Others have moved on to another area of research, called **natural-language queries.**

TIPS

The idea behind natural-language queries is that you can type a question in the same way you would ask it to a human sitting beside you — no need to keep track of Boolean operators or complex query structures. The most popular natural language query site today is AskJeeves.com, which parses the query for keywords that it then applies to the index of sites it has built. It only works with simple queries; but competition is heavy to develop a natural-language query engine that can accept a query of great complexity.

SUMMARY

- The present age is Internet age and World Wide Web is the foundation of internet evolution.
- To use the Web, in addition to an Internet connection, a user need a special piece of software called a Web browser (such as Netscape Navigator).
- We know that rules are very important to make the communication possible and TCP/IP are the protocols which make the communication possible over internet.
- Web user access information from web servers, through a client program called browser.
- The Common Gateway Interface (CGI) mechanism of HTTP enables servers to execute programs, obtain results and send the results to the requesting browser.
- Searching the World Wide Web can be both beneficial and frustrating. We may find vast amounts of information, or we may not find the kinds of information we're looking for.
- A search engine is a device that sends out inquiries to sites on the web and catalogs any web site it encounters, without evaluating it. Methods of inquiry differ from search engine to search engine, so the results reported by each one will also differ.
- A **meta-search engine** is a search tool that sends user requests to several other search engines and/or databases and aggregates the results into a single list or displays them according to their source.

- **Crawling** is the method of following links on the web to different websites, and gathering the contents of these websites for storage in the search engines databases.
- Searching through an index involves a user building a **query** and submitting it through the search engine. The query can be quite simple, a single word at minimum.

SHORT QUESTION ANSWERS

Q. 1. What is WWW?

Ans. World Wide Web is broad term, which is the collection of distributed documents referred to as "pages" located on computers (or servers) all over the world.

Q. 2. What are basics of web architecture?

Ans. The web is based on a three-part architecture :

- **HTML :** Hypertext markup language, the format from web pages, provide both formatting and hyperlinking.
- **HTTP :** Hypertext transfer protocol, the protocol for communications between Web servers and browsers.
- **CGI** – The common gateway interface, the interface for invoking programs from web servers.

Q. 3. What is URL?

Ans. URLs, or Uniform Resource Locators, are the method by which documents or data are addressed in the World Wide Web. The URL contains the following information:

- The internet name of the site containing the resource (document or data)
- The *type* of service the resource is served by (e.g., HTTP, Gopher, WAIS)
- The Internet *port number* of the service. If this is omitted the browser assumes a commonly accepted default value.
- The location of the resource in the directory structure of the server.

Q. 4. What is Web Browser?

Ans. A browser is a program that our computer runs to communicate with web servers on the Internet, which enables it to download and display the web pages that you request. Web browser understands the HTML and display text. A web browser has the ability to interpret or display many types of files.

The most popular browsers are Netscape Navigator and Microsoft Internet Explorer. Later browser is discussed in detail in succeeding sections of this unit.

Q. 5. What do you understand by the Search Engine ?

Ans. A program that searches documents for specified keywords and returns a list of the documents where the keywords were found.

Q. 6. Write some example of search engines.

Ans. Some of the example of the search engines are as follows:

1. google.com
2. lycos.com
3. altavista.com
4. hobot.com

5. excite.com
6. freefind.com
7. dogpile.com
8. infoseek.go.com
9. yahoo.com

Q. 7. What do understand by the web server? Also write some of the example of the web servers.

Ans. Web servers are a combination of software and hardware that store documents and other content. The server content is accessed using Web browsers. Web servers are available for a wide range of platforms including IBM PCs, Apple Macintosh computers, UNIX platforms and a large number of proprietary hosts.

Some the example of the web servers are as follows:

1. Tomcat
2. Weblogic
3. Apache
4. IIS

Q. 8. What do you understand by the Proxy Servers?

Ans. A proxy server (sometimes referred to as an application gateway or forwarder) is an application that mediates traffic between a protected network and the Internet. Proxies are often used instead of router-based traffic controls, to prevent traffic from passing directly between networks.

Q. 9. What do you understand by the HTTP?

Ans. HTTP is a connectionless text based protocol to make the communication possible over web. Clients (web browsers) send requests to web servers for web elements such as web pages and images. After the request is serviced by a server, the connection between client and server across the Internet is disconnected.

REVIEW QUESTIONS

1. What is WWW ?

2. Define the architecture of WWW.

3. Write the short notes on following :

(i) HTTP	(ii) TCP	(iii) IP
(iv) FTP	(v) SMTP	(vi) SSL

4. What is web browser ?

5. Write the difference between web directory and search engines.

6. Define meta search engines.

7. What are the various function of internet search engine ?

8. Write the various main component of a search engine.

❑❑❑

5

CHAPTER

HTML and XML

5.1. SGML

INSIDE THIS CHAPTER

SGML (Standard Generalized Markup Language) is an internationally agreed standard for information representation. SGML can be used for publishing in its broadest definition - from single medium conventional publishing on paper to on-line multi-media database publishing. SGML can be used to produce files which can be read by people, and exchanged between machines and applications in a straight-forward manner.

5.1.1. Features of SGML

- **SGML provides an internationally recognized, *non-proprietary* language** for writing your own markup schemes.
- **SGML relies on the principles of *descriptive markup* -** where the markup is used to indicate the nature, function or content of the data in a file, rather than saying how that data should be processed. SGML can be used to encode semantics rather than syntax. Using SGML, a heading will be identified *as* a "heading" rather than a piece of text that has to be printed or displayed in "20 point Times Bold".
- **SGML is *rigorous.*** The markup schemes that you write using SGML declare a set of rules which unambiguously state how the data must be marked up in order to be correctly structured. SGML-aware software can ensure that any markup in a file confines to the appropriate set of rules *thereby guaranteeing* that the data in that file will be structured in a *known* way. If the set of rules you are working with declares that text labeled.

- **A "sub-section" can only occur within text labeled as a "section"**, SGML-aware software will ensure that this rule is obeyed during text creation and editing.

5.2. HTML [MDU May 2005, 2006]

Definition

HTML, which stands for HyperText Markup Language, is a markup language used to create web pages. The web developer uses "HTML tags" to format different parts of the document. For example, you use HTML tags to specify headings, paragraphs, lists, tables, images and much more.

HTML is a subset of Standard Generalized Markup Language (SGML) and is specified by the World Wide Web Consortium (W3C).

HTML are initials that stand for **Hyper Text Markup Language**

- **Hyper** is the opposite of linear. It used to be that computer programs had to move in a linear fashion. This before this, this before this, and so on. **HTML** does not hold to that pattern and allows the person viewing the **W**orld **W**ide **W**eb page to go anywhere, anytime they want.
- **Text** is what you will use. English letters, just like you use everyday..
- **Markup** is what you will do. You will write in plain English and then mark up what you wrote.
- **Language.** Some may argue that technically html is a code, but you write html in plain, everyday English language.

HTML is the code behind our webpage and is what our browser looks for to display a webpage, the way the web designer intended it to look, and is a series of tags **<tags>** that tells the browser where to display what. It is really a series of simple commands that we give to the browser, just like telling our dog to sit, and because it is in plain English it is easy to learn. For example, if you want our text to show in a bold type, we command it <bold> **to be bold text** </bold>, it really is that easy.

5.2.1. What do We Need to Create HTML

We don't need any special equipment or software to create HTML. In fact, we probably have everything we need. Here is what we need :

1. **Computer**
2. **Text or HTML editor**. Most computers already have a text editor and you can easily create HTML files using a text editor. Having said that, there are definite benefits to be gained in downloading an HTML editor.

   Dreamweaver is probably the best HTML editor available, and we can download a trial version for starters. If we don't have the cash to purchase an editor, we can always download a free one. Examples include SeaMonkey, Coffee Cup (Windows) and TextPad (Windows).

   If we don't have an HTML editor, and we don't want to download one just now, a text editor is fine. Most computers already have a text editor. Examples of text editors include Notepad (for Windows), Pico (for Linux), or Simpletext/Text Edit/Text Wrangler (Mac).
3. **Web Browser**. For example, Internet Explorer or Firefox.

5.2.2. Working of HTML

HTML consists of a series of short **codes** typed into a text-file. The text is then **saved as a html file**, and **viewed through a browser**, like *Internet Explorer* or *Netscape Navigator*. This browser reads the file and translates the text into a visible form, hopefully rendering the page as the author had intended. Writing your own HTML entails using tags correctly to create your vision. You can use anything from a rudimentary text-editor to a powerful graphical editor to create HTML pages.

Here's how we look at an HTML document (known as the **"source code"**)

- When we find a page we like, click on **VIEW** at the top of the screen.
- Choose **SOURCE** from the menu.
- The **HTML** document will appear on the screen.

Try it with this page. Click on **VIEW** and then choose the **SOURCE.** It will look like mumbo jumbo right now, but soon it'll be readable and you'll be able to find exactly how a certain **HTML** presentation was performed.

5.2.3. Advantages/Disadvantages of HTML

Advantages

- Easy to use
- Loose syntax (although, being too flexible will not comply with standards)
- Supported on almost every browser, if not all browsers.
- Widely used; established on almost every website, if not all websites.
- Very similar to XML syntax, which is increasingly used for data storage
- Free - You need not buy any software
- Easy to learn and code even for novice programmers

Disadvantages

- It cannot produce dynamic output alone, since it is a static language
- Sometimes, the structuring of HTML documents is hard to grasp
- Security features offered by HTML are limited

5.2.4. HTML Document Structure **[MDU Dec. 2006]**

HTML files are just normal text files, they usually have the extension of .html, .htm or .shtml. HTML documents are generally divided into two parts, the head and the body. The body is the larger part of the document, as the body of a letter you would write to a friend would be. The head of the document contains the document's title and similar information, and the body contains the contents and everything else.

Example of a basic HTML document structure is as follows :

```
<html>
<head>
<title>Title goes here</title>
```

```
</head>
<body>
Body goes here
</body>
</html>
```

5.3. HTML TAGS

Definition

HTML works in a very simple, very logical, format. It reads like you do, top to bottom, left to right. HTML is written with TEXT. What you use to set certain sections apart as bigger text, smaller text, bold text, underlined text, is a series of tags.

If you want a line of text to be bold. You will put a tag at the exact point you want the bold lettering to start **<bold>and another tag where you want the bold lettering to stop.</bold>** There you have an example of how tags make commands and tell the browser how to display the text. Just like telling a dog to sit. You give the command.

All tags that are opened must correspondingly be closed, just as if you are writing a quoted statement with those "inverted commas". A tag is closed this way **</tag>** therefore we open a new tag **<bold> and close the tag </bold>**

Here is the summary of some basic tags :

Tags	Description
<html></html>	Tags which create the HTML document.
<head></head>	Tags which denote the first part of the HTML document.
<title></title>	Tags which define the title of the web page.
<body></body>	Tags which define the main contents of the HTML document.
<h1></h1>	Tags that indicate the headings. There are 6 headings, from h1 to h6.
<b></b>	Tags that display **bold** text.
<i></i>	Tags that display *italic* text.
<u></u>	Tags that underline the text.
 	Tag that indicate line break. No closing tag is required here.

Example 5.1

```
<html>
    <head>
    <title>My sample HTML document</title>
    </head>
    <body>
    <h1><u>My sample HTML document</u></h1>
    <h1>This is heading 1</h1>
```

```
        <h2>This is heading 2</h2>
        <h3>This is heading 3</h3>
        <h4>This is heading 4</h4>
        <h5>This is heading 5</h5>
        <h6>This is heading 6</h6>
        <b> Bold text</b>
        <br>
        <i>Italic text</i>
        <br>
        <u>Underlined text</u>
        <b><i>Bold and italic text</i></b>
        <b><i><u>Bold, italic and underlined text</u></i></b>
        <h1><i>Italic heading</i></h1>
        <h3><u>Underlined heading</u></h3>
        </body>
    </html>
```

The output of above example will be as follows:

My sample HTML document

This is heading 1

This is heading 2

This is heading 3

This is heading 4

This is heading 5

This is heading 6

Bold text
Italic text
Underlined text
Bold and italic text
Bold, italic and underlined text

Italic heading
Underlined heading

5.3.1. More HTML Tags

Now let us see some of the advance tags in following table:

TAGS	Description
<h1 align=position</h1>	Tags which aligns the heading to the left, right and the centre
<strong></strong>	Tags which strongly emphasize the bold format.
<em></em>	Tags which gives more emphasis to the italic text.
<pre></pre>	Tags which preserves the original text.
	Tags which format text as subscript.
	Tags which format text as superscript.
<hr>	Tag which insert a horizontal ruler. Closing tag is not required here.
<hr width=n%>	Tag which sets the width of the horizontal ruler.
<hr size=m>	Tag which sets the height of the horizontal ruler.
<hr nonshade>	Tag which creates the ruler without the shadow.
<p></p>	Tags which define the paragraph.
<p align=position></p>	Tags which align the paragraph to the left, right or centre.

Example 5.2

```
<html>
<head>
<title>Mr Bean</title>
</head>
<body>
<h1 align=center><u>All about  Mr. Bean</u></h1>
<hr>
<h2 align=right>Date: Nov 30,2009</h2>
<hr width=50% size=10>
<p> <strong>Mr. Bean</strong>is a 50 years old boy, currently
studying at  <em>Cambridge</em>. His  lecturer is <b>Mr. Mike
Tyson</b>. There are 33 students in his class.</p>
<p align=center> Here are some facts about Bean</p>
<pre>
                              Date of Birth      :01011900
                              Height             : 100 cm
                              Weight             : 100 kg
                              Marital Status     : Single
                              Address            : 1, Wall street,
London,UK.
</pre>
<hr nonshade>
```

```
<P align=center> If you wish to match make for Mr. Bean,
please contact him through the following phone
number:<blink>911</blink></p>
<hr>
</body>
</html>
```

The output will be as follows:

All about Mr. Bean

Date: Nov 30, 2009

Mr. Bean is a 50 years old boy, currently studying at *Cambridge*. His lecturer is **Mr. Mike Tyson**. There are 33 students in his class.

Here are some facts about Bean

Date of Birth	: 01011900
Height	: 100 cm
Weight	: 100 kg
Marital Status	: Single
Address	: 1, Wall street, London, UK.

5.3.2. Types of List

In HTML, If you wish to list out a number of items, you can use the following types of listing:

1. The Unordered List

An unordered list is a list of items. The list items are marked with bullets (typically small black circles).An unordered list starts with the <ul> tag. Each list item starts with the <li> tag. Let us discuss a example as follows:

```
<UL>
  <LH>Table Fruit</LH>
  <LI>apples
  <LI>oranges
  <LI>bananas
</UL>
```

Which could be rendered as:

Table Fruit

- apples
- oranges
- bananas

Note that <LI> tag does not require a closing tag.

Following is the example of unordered list and you can view it in your browser.

Example 5.3

```
<HTML
<HEAD><TITLE>Unordered List</TITLE></HEAD>
<BODY>
<H2> Fruits</H2>
<HR>
<UL>
<LI> Apple
<LI> Grape
<LI> Orange
<LI> Pear
<LI> Pineapple
<LI> Water Melon
</UL>
</BODY>
</HTML>
```

2. The Ordered List

Also known as the Numbered List. It is very similar in structure to the unordered list, except each list item has a number in front of it, instead of a bullet. Also, the opening tag for the list is <OL> instead of <UL>, and the closing tag is </OL> instead of </UL>. Let us see an example as follows:

```
<ol>
<li>This is List Item one ...</li>
<li>This is List Item two ...</li>
<li>This is List Item three ...</li>
</ol>
```

The output will be as follows :

1. This is List Item one ...
2. This is List Item two ...
3. This is List Item three ...

3. The Nested List

This nested list contains lists within lists. It can be clear by following example :

Example 5.4.

```
<HTML>
    <HEAD><TITLE>Nested List</TITLE></HEAD>
```

```
<BODY>
<H1 align=center><U> Internet Club</U></H1>
<OL>
<LI>Advisers
<UL>
<LI> Mr. Liew
<LI> Mrs James
<LI>Miss Irene
</UL>
<LI>Committee Members
<UL>
<LI>Nick
<LI>George
<LI>Francis
</UL>
</OL>
<BODY>
</HTML>
```

The output of above example will be as follows :

Internet Club

1. Advisers
   - Mr. Liew
   - Mrs James
   - Miss Irene
2. Committee Members
   - Nick
   - George
   - Francis

5.3.3. Adding Colors

Adding colors to your web pages could achieve certain visual effect and make them more attractive. However, you must select colors carefully and do not overuse them. You can add colour to your web page background and the text as well.

5.3.3.1. Adding Colour to the Background of the Web Page

You can specify the colour of the page's background within the body tags using the following code :

```
<body bgcolor=colour>
```

The colour attribute can be specified using normal words like red, yellow, blue etc. or by hexadecimal codes. Hexadecimal is a base 16 number system. Hexadecimal uses A for decimal 10, B for decimal 11, C for decimal 12, D for decimal 13, E for decimal 14 and F for decimal 15.

Every colour code in HTML is made up of 6 hexadecimal digits, from 000000 to FFFFFF. The hexadecimal codes might seem very complex, however, they actually make up of combination of three primary colors, i.e. red, green and blue(RGB). The last two digits specify the amount of red colour, the middle two digits specify the amount of green colour and the first two digits specify the amount of blue colour.

For example:

0000FF=maximum red 00FF00=maximum green FF0000=maximum blue.

Example 5.5

```
<html>
<head>
<title>Background Colour</title>
</head>
<body bgcolor=red>
```

The background colour of this web page is red. You can change the background colour anytime by changing the colour attribute specified by the bgcolor code. Try using the hexadecimal codes like bgcolor=FF00CC.

```
</body>
</html>
```

5.5.3.2. Adding Color to the Text

We can specify the Colour of the normal text (the default text is black) using the following code within the body tags.

```
<body bgcolor=blue text=white>
```

You may use normal words for the colour attribute or use hexadecimal notation.

Example 5.6

```
<html>
<head>
<title>Text colour</title>
</head>
<body bgcolor=blue text=FFFFCC>
```

The text colour can be formatted using the text=colour code. It is important that the colour of the text is not the same as the background colour; otherwise you won't be able to see the words.

```
</body>
</html>
```

We can format individual font colour with the Font Tag Font tag can also control font size and type of fonts.

The codes are as follows :

```
<font colour=red>Red colour text</font>
```

```
<font size=1 colour=blue>Blue text with font size 1</font>
<font type="Times New Roman">Text in Time News Roman</font>
```

Example 5.7

```
<html>
 <head>
 <title>Manipulation of Fonts</title>
 </head>
 <body bgcolor=CCFFFF  text=blue>
 <font colour=red size=4>Manipulation of </font><font
colour=brown size=4>Fonts</font>
 <hr>
 <p>
 <font colour=green type="Arial"  size=2>You can manipulate
font colour , size and font type using
 the font tags</font>.<font colour=AA0000 size=2 type="Times
New Roman">It can be done fairly
 easily by anybody. Most important you must try it out
yourself and dare to experiment with
 anything.</font>
 </p>
<hr>
This part of the text will appear blue.
 </body>
 </html>
```

5.3.4. Inserting Graphics in the Web Page

The tag for inserting graphics in a web page is

```
<img src="image filename">
```

Web page normally support two types of image file, image.gif and image.jpg. There are countless graphics files that you might copy from the World Wide Web and then insert them into your web page. You could also create your very own graphics using scanner and graphics tools like PC Paint Brush, Corel Draw, Adobe Photoshop and etc.

You could control the alignment, size of the image using the align, width and height parameter respectively as follows:

```
<img src="image.gif" align=position width=100 height=100
alt=image label>
```

The alt parameter is to show the label of the image.

Graphics could also be used as background for your web pages using the tag as follows :

```
<body background="image.gif"></body>
```

Example 5.8

```
<html>
 <head>
 <title>Inserting Graphics</title>
 </head>
 <body background=back.gif>
 <center><font size=6 colour=red>Inserting Graphics in Web
Page</font></center>
 <hr>
 <img src=image1.gif>
 <br>
 <img src=image2.gif width=100 height=100 >
  <br>
 <img src=image3.jpg width=30% height=50% align=right
alt="image3">
 </body>
 </html>
```

5.3.5. Links

Definition ***Links allow us to browse to another section of a web page, to view an in-line image, another page within a website, or another web site in the Internet.***

Let us discuss different types of links as follows :

1. Text Link

The tags are

```
<a href="filename">text</a> for linking to a local file.
<a href="URL">Web site</a> for linking to another website.
```

2. Graphical Link

The tags are just the same, but use an image in the place of text.

```
<a href="filename"><img src="image.jpg"></a>
<a href="URL"><img src="image.jpg"></a>
```

3. Email Link

This link allows users to email the person whose email address is specified in the link. The tag is

```
<a href="adeshpande@kiet.edu">Email Me</a>
```

Example 5.9

```
<html>
<head>
<title>Links
</title>
</head>
<h1>This page contains various types of link</h1>
<hr>
<body background="back.gif" link=blue alink=green vlink=red>
<a href="image1.gif">Click to view image</a>
<br>
<a href="graphic.html">Click to view this web page</a>
<br>
<a href="http://www.yahoo.com">Yahoo!, the search engine</a>
<br>
<a href="http://www.vbtutor.net">Visual Basic Tutorial</a>
<br>
<a href="image3.jpg"><img src="image2.gif"><a>
<a>
<a href="http://www.geocities.com/vkliew"><img
src="image1.gif"></a>
<hr>
<center>
<a href=" adeshpande@kiet.edu ">Email Me</a>
</body>
</html>
```

5.3.6. Introduction to frames

Definition

Frames allow us to divide the page into several rectangular areas and to display a separate document in each rectangle. Each of those rectangles is called a "frame".

Frames are very popular because they are one of the few ways to keep part of the page stationary while other parts change.

These properties make possible:

- **Static frames :** Elements that a user should always see, such as button bars, logos, copyright notices, and title graphics, can be placed in individual frames that are locked into place on the user agent window.
- **Live frames :** Documents, icons, interactive forms, videos, multimedia, topic maps, and anything else that can react to user input or programmed activity.

- As a user navigates a site in "live" frames, the contents of static frames remain fixed, even though adjoining frames redraw.
- **Functional tables of contents :** A frame can contain interactive Tables of Contents (TOCs) with links that, when clicked, display results in an adjoining frame. These TOCs can be static or interactive with collapsible lists, graphical maps of document structure, or displays of file and link architectures.
- **Single-page query and answer displays :** Frames designed side-by-side permit queries to be posed and answered on the same page, with one frame holding the query form, and the other presenting the results.
- Basically any two-dimensional arrangement of static and interactive frames that you want. You can place information in any rectangular block independently of the overall page or of other frames on the page.

5.3.6.1. Frame Documents

Frames divide a browser window into two or more document windows, each displaying a different document, or a different part of the same document. Frames in an HTML document can cause a web page to appear to be divided into multiple, scrollable regions. For each frame, you can assign a name, a source document locator, dimensions, border alignment and decorations, scroll and resize behaviors, loading and unloading behavior, file and topic maps, and style sheets.

- **Names :** You can place an anchor in any frame, link to any addressable object, and place the object into any named frame.
- **Source document locator :** You can use whatever addressing schemes your user agent supports, including URLs and filenames.
- **Dimensions :** You can rigidly or flexibly layout a two-dimensional grid of rectangular blocks.
- **Border alignment and decoration :** You can adjust the position of the left and right margins, the top and bottom margins, and the alignment of the frame. You can also make the borders of a frame invisible.
- **Scrolling :** Frames can have scrollbars, no scrollbars, or you can let the browser turn them on if the document is larger than the current horizontal or vertical size of the frame.
- **Resizable :** Frames are normaly resizable in the browser, but that can be disabled so the frame may not be resized at the user agent.
- **Loading and unloading :** You can provide a script to be run when the user agent finishes loading all frames or when the user exit the document.
- **File and topic maps :** You can place a file or topic navigator into a frame. The navigator might be a collapsible listing of file system, a listing of document headings, thumbnails of images in a document, or an index of any element type.

5.3.6.2. Frame Document Structure

Frames in HTML documents are created and controlled through the structure of three element types: FRAMESET, FRAME and NOFRAMES.

A web page containing frames is created by a main or "framing" document. This document defines the frame regions on the user agent's page or window, and addresses the documents or objects that initially appear in the frames. Hypertext anchors within any frame can also target specific frames to place the contents of addressable documents and objects.

The structure of a frame-enabling HTML document type is similar to usual HTML structure. The key difference in an HTML document type is that the BODY container element is basically replaced by a FRAMESET container element. The initial FRAMESET element describes the frames that make up the page, and the FRAME elements specify the sub-documents that appear initially in each.

A FRAMESET may only contain nested FRAMESET and FRAME elements. When a FRAMESET element is nested within another FRAMESET element, it acts as a subframe. A BODY element can follow the FRAMESET to provide an alternative document for user agents that do not support frames. And a BODY element can contain NOFRAMES elements whose contents are not to be rendered by user agents which support frames.

A frame document contains a set of frames, each of which likely uses the BODY element. Elements that might normally be placed in the BODY element should not appear before the first FRAMESET element, or the FRAMESET will be ignored. The FRAMESET tag has a matching end tag.

5.3.6.3. Frame Document Syntax

Frame syntax is similar in scope and complexity to that used by tables, and has been designed to be quickly processed by Internet client layout engines.

```
<HTML>
<HEAD>
</HEAD>

<FRAMESET rows="33%,33%,33%">
  <FRAMESET cols="50%,50%">
      <FRAME name="frame1">
      <FRAME name="frame2">
  </FRAMESET>
  <FRAME name="frame3">
  <FRAME name="frame4">
</FRAMESET>
<BODY>
...contents to display in non-frame-capable user agent...
</BODY>

</HTML>
```

5.3.6.4. Basic Frame Concepts

Try to remember the following basic frameset concepts:

- The opening FRAMESET tag defines the actual size and layout of the frames.
- Using the ROWS and COLS attributes of the FRAMESET tag, you define frames, then **fill** them with the FRAME element, or **subdivide** them with nested FRAMESET elements.

- The number of entries in ROWS or COLS attribute values determine how many frames are displayed and what size they are. If you have both ROWS and COLS, multiply the number of entries in each attribute value to calculate the total number of frames (i.e., three entries in the ROWS attribute and four entries in the COLS attribute create a grid of twelve frames).
- The FRAME elements fill the frames defined by FRAMESET, and are applied in the frame set in a left-to-right, top-to-bottom order.
- FRAME elements which include the SRC attribute, define the contents of each frame you create.
- Frames are **always** rectangular.

5.3.6.5. The FRAME Element

The FRAME element defines a single frame in a frameset. It has 7 possible attributes: SRC, NAME, FRAMEBORDER, MARGINWIDTH, MARGINHEIGHT, SCROLLING, and NORESIZE. The FRAME tag is not a container so it has no matching end tag.

The attributes of the FRAME element type are as follows:

src="address"

Specifies the address of the document to be displayed in the frame. When omitted the frame is displayed as a blank space.

name="window_name"

The NAME attribute is used to assign a name to a frame so it can be targeted by links in other documents. (These are usually from other frames in the same document.) The NAME attribute is optional; by default all windows are unnamed. Names must begin with an alphabetic character. Named frames can have their window contents targeted with the TARGET attribute.

frameborder="1|0"

FRAMEBORDER is a boolean which triggers the display of frame separators around the frame. When set to "1" a separator is drawn on every side next to another frame. When set to "0" the decision to draw separators is left to surrounding frames. Indeed in this case separators might still be drawn because FRAMEBORDER is set "1" on adjacent frames. Default is "1".

marginwidth="value"

The MARGINWIDTH attribute is used when the document author wants some control of the margins for this frame. If specified, the value for MARGINWIDTH is in pixels. Margins cannot be less than one; this insures that frame objects won't touch frame edges, and can't be specified in a way that leaves no space for the document contents. The MARGINWIDTH attribute is optional; by default, all frames default to letting the browser decide on an appropriate margin width.

marginheight="value"

The MARGINHEIGHT attribute is just like MARGINWIDTH above, except it controls the upper and lower margins instead of the left and right margins.

noresize

The NORESIZE attribute doesn't require a value. It can simply be used as a flag to indicates that the frame is not resizable by the user. Users typically resize frames by dragging a frame edge to a new position. Note that if any frame adjacent to an edge is not resizable, that entire edge is restricted from moving. This affects the resizability of other frames. The NORESIZE attribute is optional; by default all frames are resizable.

scrolling="yes|no|auto"

The SCROLLING attribute indicates whether the frame should have scrollbars or not. "yes" results in scrollbars always being visible on that frame. "no" results in scrollbars never being visible. "auto" instructs the browser to decide whether scrollbars are needed, and to place them where necessary. That is, with "auto", the frame has scrollbars only if the document is larger than the current size of the frame. The SCROLLING attribute is optional; the default value is "auto."

Example : This small document causes three frames to be displayed. The first (outer) FRAMESET element divides the browser window into two horizontal regions. The upper region contains a single frame displaying the document **http://www.zenith.com/**, and the lower region displays the frames defined in the second (inner) FRAMESET element. This FRAMESET divides the lower part of the window into two vertical frames, displaying the documents **http://www.sinister.org/** and **http://www.dexter.org/**.

```
<HTML>
<FRAMESET rows="40%,60%">
    <FRAME src="http://www.zenith.com/">
    <FRAMESET cols="50%,50%">
        <FRAME src="http://www.sinister.org/">
        <FRAME src="http://www.dexter.org/">
    </FRAMESET>
</FRAMESET>
</HTML>
```

5.3.6.6. Setting Frame Width and Height

There are three ways to specify the width or height of a frame:

- As a percentage of the area allotted to the parent frame set.
- As an absolute (specific) number of screen pixels (e.g., 250).
- As a 'relative size'.

Example for 3 rows, the first and the last being smaller than the center row:

```
<FRAMESET rows="20%,60%,20%">
```

Example for 3 rows, the first and the last being fixed height, with the remaining space assigned to the middle row:

```
<FRAMESET rows="100,*,100">
```

A 'relative size' is specified with an asterisk, e.g., '1*', '2*', '3*' ('1*' can also be written simply as '*'). This is interpreted as follows: after all widths (or heights) specified as percentages or absolute amounts have been allocated to the corresponding frames, the remaining space is allocated to frames whose widths (or heights) have been specified as a relative size. The amount of space allocated to a frame is proportional to the number in front of the asterisk. For example:

```
<FRAMESET rows="30%,400,*,2*">
    <FRAME ... >
    <FRAME ... >
    <FRAME ... >
    <FRAME ... >
</FRAMESET>
```

Suppose the browser window is currently 1000 pixels high. The first frame gets 30% of the total height, that is, 300 pixels; the second frame gets 400 pixels, since an absolute amount was specified. This leaves 300 pixels to be divided between the other two frames. The fourth frame's height is specified as '2*', so it is twice as high as the third frame, whose height is only '*' (1*). Therefore the third frame is 100 pixels high and the fourth is 200 pixels high.

Examples 5.10. This example shows the HTML source to achieve the layout below using frames.

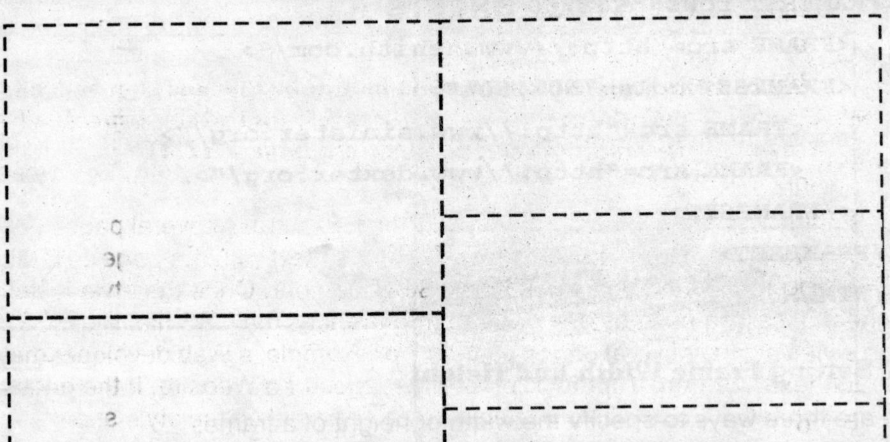

```
<FRAMESET cols="50%,50%">
    <FRAMESET rows="50%,50%">
        <FRAME src="cell.html">
        <FRAME src="cell.html">
    </FRAMESET>
```

```
        <FRAMESET>
            <FRAME src="cell.html">
            <FRAME src="cell.html">
            <FRAME src="cell.html">
        </FRAMESET>
    </FRAMESET>

    <BODY>
    <H1>Frame Alert!</H1>
    <P>
```

This document is designed to be viewed using the frames feature. If you see this message, you are using a non-frame-capable web client.

```
    </P>
    </BODY>
```

The main layout in two columns is obtained by nesting two framesets inside the topmost frameset. The sub framesets then respectively contain two and three frames. At the end a BODY element is used to provide an alternate document for non frame based renderings.

5.4. CASCADING STYLE SHEETS (CSS) [MDU May 2008]

Definition ***Cascading style sheets are used to format the layout of Web pages. They can be used to define text styles, table sizes, and other aspects of Web pages that previously could only be defined in a page's HTML.***

CSS helps Web developers create a uniform look across several pages of a Website. Instead of defining the style of each table and each block of text within a page's HTML, commonly used styles need to be defined only once in a CSS document. Once the style is defined in cascading style sheet, it can be used by any page that references the CSS file. Plus, CSS makes it easy to change styles across several pages at once. For example, a Web developer may want to increase the default text size from 10pt to 12pt for fifty pages of a Website. If the pages all reference the same style sheet, the text size only needs to be changed on the style sheet and all the pages will show the larger text.

5.4.1. Advantages of CSS

CSS stands for Cascading Style Sheet. CSS is used to improve the presentation of the content within HTML tags. Using CSS we can define different ways of content display based on the media.

We will discuss some advantages of using CSS as follows:

1. Less code : As we define the property with value at a common style class and use the same class at different locations, so we use less code. Less code means less bandwidth consumption and easy to maintain the content.

2. Easy maintenance : As we define styles globally or at a common place so any changes became easy. For example in a site we have all the name of the products, displayed using a particular style property. Now by changing the style class at external style sheet for the product name we can change the style through out the site. We can keep more than one style sheet and use them based on requirement. With property inheritance methods maintaining of different styles of the same tag became easy.

3. Higher content to code (tag) ratio : We can achieve higher content to code ratio in a page by using CSS as we can shift the style declarations to an external file. This is important for search engine point of view. We present less tags (style tags) and more content to the spiders for indexing.

4. First download of pages : As browsers cache the style sheet page so the page loading became fast. The styles classes are not loaded from server each time different pages of same site using same CSS is used.

5. Flexibility in defining style : The name cascading indicates that we can use more than one style and the priority is given to local styles first. We can override the global style declared and locally assign the style to the tag.

5.4.2. How To Use CSS

One thing that CSS allows us to use for screen presentation are alternate cursors. This is not the idea of downloading or forcing a download of a cursor, as was done in the past (though that is possible as well), but instead, we use several built in concept cursors.

Each of these cursors you can use to enhance useability of your website. For instance, if you want to define that something has context help, you could use the help cursor.

The CSS2 standard gives us 17 options for cursors, plus the option to use an external cursor from a URI. For two of the cursor options, there could have been slightly better phrasing, because it is a touch confusing for some people to remember. Two of the options are auto and default.

Let us see some of the cursors as follows :

1. auto - Initial value. Lets the browser choose.
2. default - Displays the basic cursor (usually an arrow).
3. crosshair - A '+' shaped cursor.
4. pointer - A pointer (usally displayed for links).
5. move - Indicates a moveable element (often a 4-way arrow)
6. text - Text can be selected. (often an I-beam)
7. wait - Asks the user to wait (often an hourglass).
8. help - Indicates help is available (often a question mark).
9. progress - Similar to wait but shows that a process is working.
10. e-resize - Indicates a resizeable element. Double-arrows display direction to size.
11. ne-resize - Same as e-resize.
12. nw-resize - Same as e-resize.
13. n-resize - Same as e-resize.
14. se-resize - Same as e-resize.
15. sw-resize - Same as e-resize.
16. s-resize - Same as e-resize.
17. w-resize - Same as e-resize.

We mentioned that there is also the option to use the cursor for a call to a remote cursor. So, let's say that you have a cursor you would like the browser to use, and you have it stored at a specific location. You could do the following:

```
<p style="cursor: url(preferredcursor.cur)">Text
```

Now, the problem is that not all browsers may be able to load or use that cursor. Let me give an example of one that browsers might have a hard time viewing, and the way you can still set an alternative.

In CSS or Inline: acroynm { cursor: url(1.svg), url(2.cur), help }

The first one (1.svg) is the preferred cursor and if the browser can handle loading and displaying and SVG file, it will do that one. If not, it moves to the second, and tries to load the 2.cur file as the cursor. If it cannot, it will instead display the built-in help cursor.

That is the nice aspect of CSS cursors - they degrade gracefully. If a browser cannot support them, they don't show, and it does not cause display problems. So, feel free to use the CSS cursors. Not all will show for everyone, but due to the smooth degradation for browsers without support - it is a perfect use of CSS.

5.5. LISTING OF HTML TAGS

Below are the most common used html tags used in website design and a short description of the function of each. Some have examples of how they are used with their html codes. Website HTML tags are arranged in alphabetical order.

1. <A></A> This element is what the hyperlinked structure of the World Wide Web is based on. It is used in two ways:
   (i) Create a hyperlink to another anchor (Link to another page or website)
   (ii) Create an anchor in a document .
2. <ABBR></ABBR> Identifies that the content is an abbreviated form of some kind.
3. <ACRONYM></ACRONYM> Identifies that some text is an acronym.
4. <ADDRESS></ADDRESS> Specifies information such as authorship and contact details for the current document. Browsers should render the content with paragraph-breaks before and after.
5. <APPLET></APPLET> Embed a Java applet into your website document.
6. <AREA></AREA> Used to implement a client-side image map.
7. <AUDIOSCOPE> This element displays the audioscope, the graphical display of the amplitude of the current sound over time.
8. <B></B> Change your website text to bold. <bold></bold> or <strong></strong> can also be used.
9. <BASE> Supply a base address that must be used for resolving relative URI's to absolute URI's.
10. <BASEFONT> Change the appearance of the default font that is used to draw the text.
11. <BDO></BDO> This element overrides the bidirectional algorithm, the default algorithm to resolve the direction to show the text.
12. <BGSOUND> Play a background sound when your webpage is opened. This element must be placed in the HEAD section of the document.

13. <BIG></BIG> Increase the current size of the font by 1. The maximum size is 7.
14. <BLACKFACE></BLACKFACE> This element will render the text in a double-weight boldface font.
15. <BLINK></BLINK> Changes the text to blinking. Drives people insane. *smiles*
16. <BLOCKQUOTE></BLOCKQUOTE> This is used to enclose larger quotations from other works in the page.
17. <BODY></BODY> This element contains the body of your website document. If a <HEAD> section is present in the page then the body must be placed after this section.
18. <BQ></BQ> This element is an alias of the blockquote element.
19.
 Line Break. Break the current line and continue on the next line.
20. <BUTTON></BUTTON> This attribute creates a button that the user can push.
21. <CAPTION></CAPTION> Specify the caption of a table. This element is only valid inside the TABLE element.
22. <CENTER></CENTER> Centers everything inside the opening and closing element. This tag is equivalent to <DIV align="center">.
23. <CITE></CITE> Used for citations or references to other sources.
24. <CODE></CODE> Used for source code examples.
25. <COL> This element sets the attribute values for one or more columns.
26. <COLGROUP> This tag creates a column group and sets attribute values for all the columns in this group.
27. <COMMENT></COMMENT> Used to insert comments in the HTML source which will be ignored by the browser. All HTML elements inside the comment will be ignored. You can use this code to put comments in your pages, which can help you when you have to edit the source later.

    HTML generating programs sometimes store program-specific information inside comments, so they will not be visible, but still available to the program. This code is not a container, but inside it you can put one or more comments, by surrounding with "—". The end of the code is indicated with the sequence —>.
28. <DD></DD> The description of a term in a definition list.
29. <DEL></DEL> This is a tag that is used to indicate web page text that has been deleted.
30. <DFN></DFN> This is a element that is used to indicate a word or phrase that is being defined.
31. <DIR></DIR> Create a directory list..
32. <DIV></DIV> HTML DIV tags. This element is a general container for a part of the contents of a page. Through the DIV element you can add attributes, like style information, to this whole division. The DIV element will not show anything when used without any other attribute. A division will terminate a paragraph opened with the P element. <DIV align="center"> is the same as the deprecated <CENTER> element.
33. <DL></DL> Create a definition list. This is a list where each item consists of two parts.
34. <EM></EM> Put "*emphasis*" on the enclosed text.
35. <EMBED></EMBED> The EMBED element lets you display output from a plug-in application in an HTML document.

36. <FIELDSET></FIELDSET> Group a set of related controls in a form together.
37. <FONT></FONT> HTML font tags change the font which is used to draw the text.
38. <FORM></FORM> HTML form tags. Create a form inside a document.
39. <FRAME> In a page with frames, this element defines how a specific frame looks and what is initially shown inside the frame, when the framed page is loaded. This element is only allowed inside a FRAMESET element.
40. <FRAMESET></FRAMESET> Container for creating a document that consists of several frames.
41. <H1></H1> The elements H1, H2, H3, H4, H5 and H6 are used to create several levels of headers, with H1 as the most important header and H6 as the least important.
42. <HEAD></HEAD> Container for elements describing the current document. This section contains no contents the browser should display in the body of the text. The following elements are allowed inside the HEAD section :

    BASE
    BASEFONT
    BGSOUND
    ISINDEX
    LINK
    META
    SCRIPT
    STYLE
    TITLE

43. <HR> Draw a horizontal rule.
44. <HTML></HTML> The container for a complete HTML document.
45. <I></I> Change the text to italic. <italic></italic> can also be used.
46. <IFRAME></IFRAME> This element is a container to create an inline or floating frame. A floating frame is a frame in which the contents of another HTML document can be seen.
47. <ILAYER></ILAYER> With this element you can create several layers of content on a page. These layers can be stacked on top of each other, showing parts of underlying layers through non-occupied space.
48. <IMG> HTML image tag. Place an image in the document.
49. <INPUT> Create a control for a form. A control is an element which the user can use to enter data, like textboxes, radiobuttons and checkboxes and is only valid inside the FORM element.
50. <ISINDEX> This element will show an text input field. After pressing the <Enter> key the browser will construct a new URL, with the current address, a question mark and the text the user entered in the text field, and send it to the server
51. <KBD></KBD> This is used to identify text that a user is supposed to enter.
52. <LABEL></LABEL> Attach information to a specific field of a form.
53. <LAYER></LAYER> With this element you can create several layers of content on a page. These layers can be stacked on top of each other, showing parts of underlying layers through non-occupied space.

54. <LEGEND></LEGEND> Give the caption for a group of related controls, created with the FIELDSET element.
55. <LI></LI>List tags Identify an item in a list..
56. <MAP></MAP> This element is a container for the map that is used in a client-side image map.
57. <MARQUEE></MARQUEE> HTML marquee tag. This element is a container that enables you to create a scrolling text marquee.
58. <MENU></MENU> A container for a list of menu items.
59. <META> This element supplies meta-information about the current document.
60. <NOBR></NOBR> The NOBR element stands for NO BReak. This means all the text between the start and end of the NOBR elements cannot have line breaks inserted between them.
61. <NOEMBED></NOEMBED> This element defines content within EMBED content that is to be ignored by browsers that can activate the EMBED plug-in application.
62. <NOFRAMES></NOFRAMES> This element provides a way to create alternative content that is intended for browsers that can't show frames, or are configured not to show them. A browser that displays the frames ignores the contents of the NOFRAMES element.
63. <OBJECT></OBJECT> The object element allows the author to embed an object into the document. This element also replaces the APPLET element.
64. <OPTION></OPTION> This describes an option in a listbox of a form.
65. <P></P> This indicates a paragraph in the document. It is a container but most browsers allow you to omit the closing element.
66. <PARAM> This element is for supplying parameters to a JAVA applet or another object and is only valid inside the APPLET and OBJECT elements.
67. <PLAINTEXT></PLAINTEXT> All HTML elements inside this container are ignored by the browser, and shown as they were only text.
68. <PRE></PRE> This element allows you to show preformatted text as it is, using the supplied whitespace of the text.
69. <Q></Q> This is used to enclose short quotations from other works in the page.
70. <S></S> Render text as strikethrough.
71. <SAMP></SAMP> This element describes text that is output from a program.
72. <SCRIPT></SCRIPT> This element adds the possibility of programming inside a HTML document by using a scripting language.
73. <SELECT></SELECT> This element lets you create a listbox as an input field on a form.
74. <SERVER></SERVER> This element is used to write JavaScripts that will be executed on the server, in the process of serving the page to the browser.
75. <SMALL></SMALL> Draw the text using a smaller font than the one that is used for normal text.
76. <SPACER> With a spacer you can control the horizontal whitespace that appears between words in a line, the vertical whitespace that appears between lines on a page, or set up rectangular spacing elements.

77. <SPAN></SPAN> This element is used to create a structure in a document. By using this element you can give a part of the document a name, or apply style sheet information to the part.

78. <STRIKE></STRIKE> Render text as strikethrough. Same as the <S> tag.

79. <STRONG></STRONG> Render the text with strong emphasis.

80. <STYLE></STYLE> This element is a container for style sheet elements to use with this document.

81. This is a container for text that should be displayed as a subscript, and, if practical, using a smaller font (compared with normal text).

82. Display the text as a superscript.

83. <TABLE></TABLE>

    HTML table tags. Create a table layout which can contain cells in rows and columns. The cells of a table are specified with the TR, TH and TD elements.

84. <TBODY></TBODY>

    This element defines the body part of a table.

85. <TD></TD>

    HTML table tags. The container for a cell in a table. Inside this you can put all the HTML coding you want to appear in the cell.

86. <TEXTAREA></TEXTAREA>

    Define a multiline text field in a form. This element is only valid inside the FORM element.

87. <TFOOT></TFOOT>

    This element defines the footer of a table.

88. <TITLE></TITLE>

    Specify the title of the HTML document. This element is only allowed inside the HEAD element.

89. <TR></TR>

    Define a row inside a table.

90. <U></U>

    Underline your text.

91. <UL></UL>

    Create an unordered list of items, where unordered means the individual items are not numbered, but have a bullet in front of them. If you want numbered items use an ordered list. The items in the list are identified with the LI element.

92. <VAR></VAR>

    Used to describe a metasyntactic variable, where the user is to replace the variable with a specific instance. Typically displayed in italics.

93. <WBR>

    The WBR element stands for Word Break.

5.6. XML

[MDU May 2005, 2006, 2008]

Because of the lack of SGML support in mainstream Web browsers, most applications that deliver SGML information over the Web convert the SGML to HTML. This down-translation removes much of the intelligence of the original SGML information. That lost intelligence virtually eliminates information flexibility and poses a significant barrier to reuse, interchange, and automation.

For this reason, XML (Extensible Markup Language) was developed by the XML working group (known as the SGML Editorial Review Board) formed under the auspices of the World Wide Web Consortium (W3C) in 1996.

XML is a highly functional subset of SGML. The purpose of XML is to specify an SGML subset that works very well for delivering SGML information over the Web. When the mainstream Web browsers support XML, it is believed that it's going to be very easy to publish SGML information on the Web.

Definition

XML stands for Extensible Markup Language and it is a markup language much like HTML. It is a meta language to let users design their own markup language and was designed to carry data, not to display data.XML tags are not predefined and we must define our own tags. XML is designed to be self-descriptive

5.6.1. Features/Benefits of XML

XML has been widely adopted since its creation and with good reason. Some of the key features and benefits of XML include:

1. **Easy data exchange :** One of the great things about XML is that it can allow easy sharing of data between different applications - even if these applications are written in different languages and reside on different platforms.
2. **Self-describing data :** When you look at an XML document, it is very easy to figure out what's going on.
3. **Create your own languages :** XML allows you to specify your own markup language for your own specific purpose. Some existing XML based languages include Banking Industry Technology Secretariat (BITS), Bank Internet Payment System (BIPS), Financial Exchange (IFX) and many more.

5.6.2. The Difference Between XML and HTML

XML looks similar to HTML. Like XML, HTML is also a markup language. In fact, HTML stands for Hypertext Markup Language. Markup languages are used for describing how a document's contents should be interpreted.

XML is not a replacement for HTML.

XML and HTML were designed with different goals :

1. XML was designed to transport and store data, with focus on what data is.
2. HTML was designed to display data, with focus on how data looks.

HTML is about displaying information, while XML is about carrying information.

5.6.3. Use of XML

XML is used in many aspects of web development, often to simplify data storage and sharing.

1. XML Separates Data from HTML

If you need to display dynamic data in your HTML document, it will take a lot of work to edit the HTML each time the data changes. With XML, data can be stored in separate XML files. This way you can concentrate on using HTML for layout and display, and be sure that changes in the underlying data will not require any changes to the HTML.

2. XML Simplifies Data Sharing

In the real world, computer systems and databases contain data in incompatible formats.XML data is stored in plain text format. This provides a software- and hardware-independent way of storing data. This makes it much easier to create data that different applications can share.

3. XML Simplifies Data Transport

With XML, data can easily be exchanged between incompatible systems. One of the most time-consuming challenges for developers is to exchange data between incompatible systems over the Internet. Exchanging data as XML greatly reduces this complexity, since the data can be read by different incompatible applications.

4. XML Simplifies Platform Changes

Upgrading to new systems (hardware or software platforms), is always very time consuming. Large amounts of data must be converted and incompatible data is often lost.XML data is stored in text format. This makes it easier to expand or upgrade to new operating systems, new applications, or new browsers, without losing data.

5. XML Makes Your Data More Available

Since XML is independent of hardware, software and application, XML can make your data more available and useful. Different applications can access your data, not only in HTML pages, but also from XML data sources. With XML, your data can be available to all kinds of "reading machines" (Handheld computers, voice machines, news feeds, etc), and make it more available for blind people, or people with other disabilities.

6. XML is Used to Create New Internet Languages

A lot of new Internet languages are created with XML. Here are some examples:

(a) XHTML the latest version of HTML
(b) WSDL for describing available web services
(c) WAP and WML as markup languages for handheld devices
(d) RSS languages for news feeds
(e) RDF and OWL for describing resources and ontology
(f) SMIL for describing multimedia for the web

5.6.4. What Does XML Look Like?

The following example demonstrates what the contents of a typical XML document could look like.

```
<?xml version="1.0" encoding="UTF-8"?>
<tutorials>
  <tutorial>
    <name>XML Tutorial</name>
    <url>http://www.quackit.com/xml/tutorial</url>
  </tutorial>
  <tutorial>
    <name>HTML Tutorial</name>
    <url>http://www.quackit.com/html/tutorial</url>
  </tutorial>
</tutorials>
```

5.6.5. XML Viewers

While you can view XML documents with a simple text editor, there are free XML viewers that present the contents in a more readable form. Most modern browsers include XML support, so it's quite possible that your own browser is able to display the contents of XML files.

TIPS

You open an XML file in your browser the same way you open any other file in your browser. If it's a local file you can type the full path into the address bar. Otherwise, if it's available over the web, you can type the URL into the address bar.

1. Notepad

You can use a text editor such as Notepad to create or view a simple XML file. Here's what this XML file looks like in Notepad:

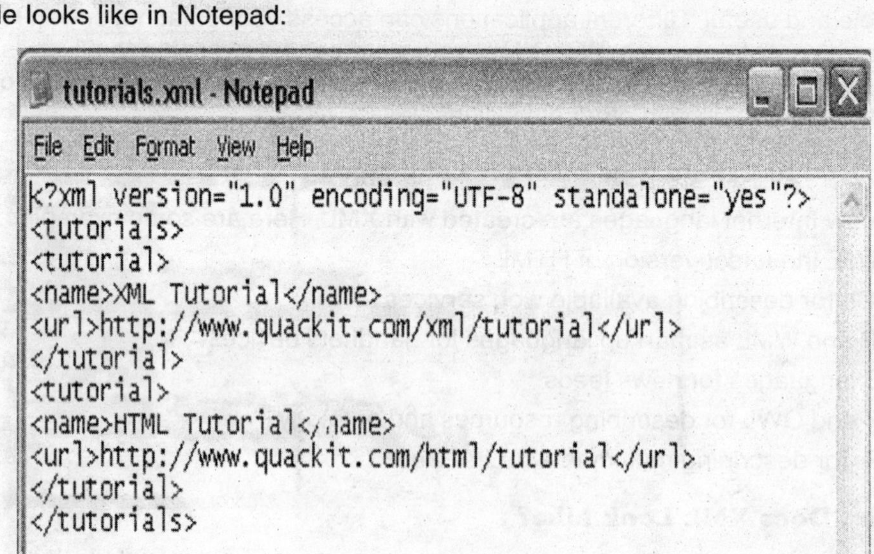

2. Firefox

Here's how the above XML file appears in Firefox 2.0. Notice the difference between Notepad and Firefox. Firefox actually attempts to interpret the document. It even warns us if it thinks something could be out of place (such as a missing stylesheet).

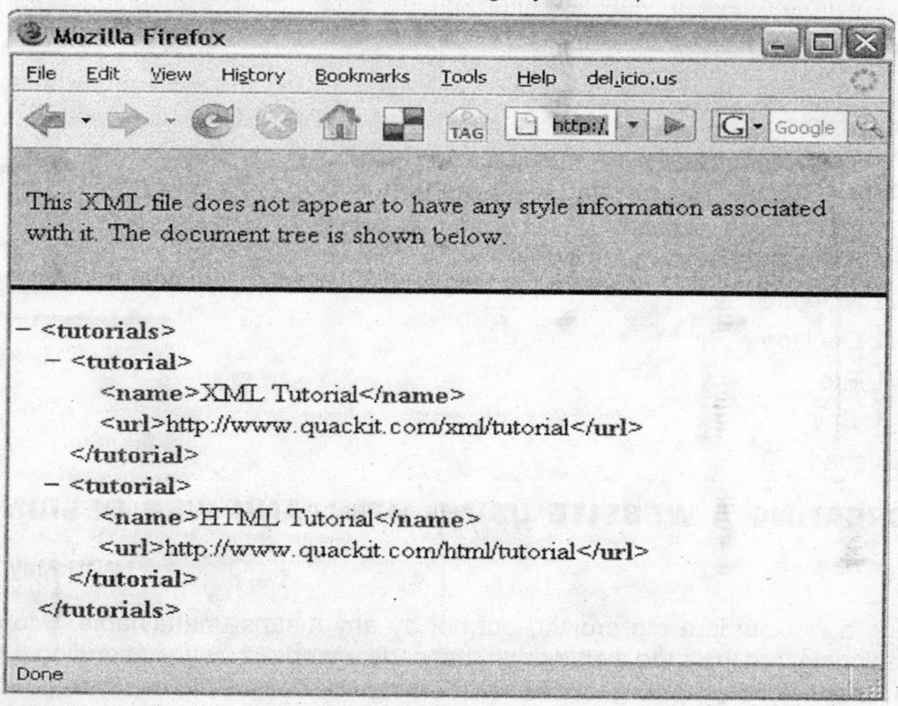

3. Internet Explorer

Let us see, how the same XML file appears in Internet Explorer 6.0

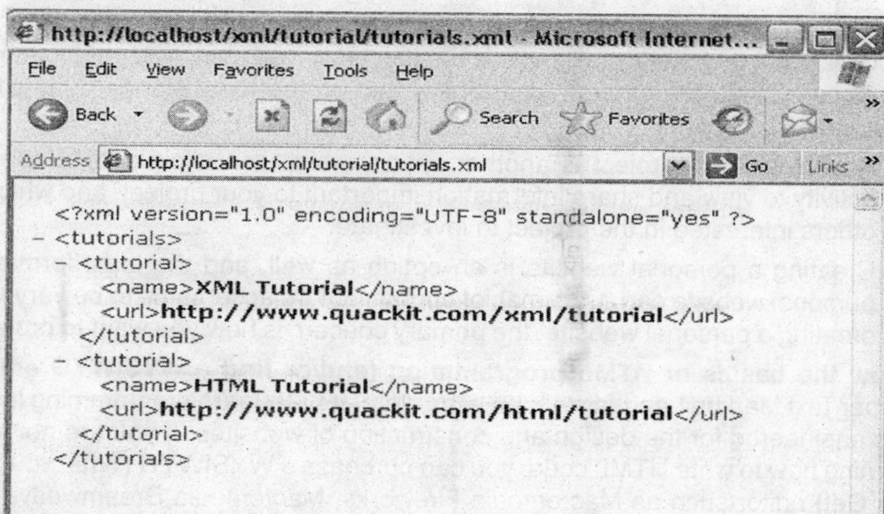

5.6.6. XML Editors

[MDU May 2005]

We can create XML documents using a simple text editor such as Notepad, WordPad, vi, emacs, or SimpleText etc. Text editors are OK, but if you're serious about creating XML, you'd be better off finding an XML editor.

Benefits of an XML Editor

An XML editor will make your life easier when coding XML and more importantly, it will help prevent you from making errors.XML editors are similar to HTML editors (or any other programming editor) in that they provide syntax highlighting which helps with readability when you have coding. They'll also do things like automatically insert a closing tag once you have added an opening tag.

A good XML editor should also provide validation - a way for you to validate that the documents you create are well formed.Here are some XML editors you might find useful:

1. XML Notepad
2. XML Cooktop
3. XML Pro
4. XML Spy
5. eNotepad

5.7. CREATING A WEBSITE USING HTML AND WEB DESIGN

[MDU May 2004, 2006]

Creating a website is a big project, but not by any means unattainable. More and more people are finding that they, too, can design and code a website by just spending a few hours in study and deciding on specific goals. The following will outline the basic steps in creating a website.

1. **Determine the purpose of your website.** It is important that you have a clear goal in mind when creating a website, even if that goal is as simple as creating a place to display your stuff.
   - Creating a professional website is the most involved choice, requiring more care and planning than the others; accessibility and professional standards must be kept in mind.
   - A website for a project is another option, which can allow people specific to one activity to view and share information important to your project, and which can allow others interested in the project to investigate.
   - Creating a personal website is an option as well, and the least formal one. Your personal website can be formal, of course, but it's acceptable to be very laid back. In creating a personal website, the primary concern is how you want to project yourself.
2. **Know the basics of HTML programming (and/or find a WYSIWYG editor).** HTML (Hyper Text Markup Language) is an extremely easy-to-learn programming language that was engineered for the design and construction of websites. If you are not interested in learning how to write HTML code, you can purchase a WYSIWYG (What You See Is What You Get) editor such as Macromedia Fireworks, Macromedia Dreamweaver (an earlier version), or Microsoft FrontPage.
3. **Consider CSS.** CSS stands for Cascading Style Sheets, and is a useful way of storing modifications to the *style* of your website in an external document which is referenced for

the display of our website. It is not necessary to the first steps in web design, but it is efficient, powerful, and in some time may replace many of the style elements used in HTML. It will also help you to separate the content and the design of your website into more manageable chunks.

4. **Decide how you're going to handle different browser capabilities.** You may have noticed over your time surfing the internet that some webpages look great in Internet Explorer, but suffer mutilation in Netscape (Mozilla and Opera, the other two major contenders, have quirks of their own). One of the reasons for this is the difference in the way that each browser handles incorrect code. If you were to view a webpage with a missing </td> tag or table tags in the wrong order, Netscape would display a disaster, while Internet Explorer would intuitively "fix" your tags for you and display a working table. Some believe this is virtue on the part of IE, while others believe that a browser should not guess what you're trying to do.
5. **Know your aesthetic capabilities.** Some people find web design a useful outlet for their artistic talent, and others are more comfortable creating an organized, code-based look for their website (which is creative in its own right). If you enjoy the idea of designing graphics for use on your website, then go for it. If not, then don't feel limited. In a knowledge of programming, there is incredible potential to make your website both functional and good looking. Experiment and branch into other types of web-building technologies (such as CSS, JavaScript, and XHTML) to further expand the possibilities.

After you have completed these steps, the last thing to do is write the code for your website. After you have successfully designed your website, purchase a domain name, find webhosting, and you're on the air.

SUMMARY

- Website designing is very important activity and HTML is very basic and widely used tool for it. This chapter will give basic working knowledge of HTML and XML to the students.
- HTML files are just normal text files, they usually have the extension of .html, .htm or .shtml.
- HTML Frames allow you to divide the page into several rectangular areas and to display a separate document in each rectangle. Each of those rectangles is called a "frame".
- XML stands for Extensible Markup Language and it is a markup language much like HTML.XML provides the facility to carry data.
- XML is a highly functional subset of SGML. The purpose of XML is to specify an SGML subset that works very well for delivering SGML information over the Web.

SHORT QUESTION ANSWERS

Q. 1. What is SGML?

Ans. SGML (Standard Generalized Markup Language) is an internationally agreed standard for information representation.

Q. 2. What do you understand by HTML?

Ans. **HTML,** which stands for HyperText Markup Language, is a markup language used to create web pages. The web developer uses "HTML tags" to format different parts of the document. For example, you use HTML tags to specify headings, paragraphs, lists, tables, images and much more.

Q. 3. Define HTML Tags?

Ans. **HTML** works in a very simple, very logical, format. It reads like you do, top to bottom, left to right. **HTML** is written with **TEXT**. What you use to set certain sections apart as bigger text, smaller text, bold text, underlined text, is a series of tags.

Q. 4. What is Unordered List in HTML?

Ans. An unordered list is a list of items. The list items are marked with bullets (typically small black circles). An unordered list starts with the <ul> tag. Each list item starts with the <li> tag.

Q. 5. What is frame in HTML?

Ans. Frames allow you to divide the page into several rectangular areas and to display a separate document in each rectangle. Each of those rectangles is called a "frame". Frames are very popular because they are one of the few ways to keep part of the page stationary while other parts change. Frames are also one of the most controversial uses of HTML, because of the way the frames concept was designed, and because many web framed web sites are poorly implemented.

Q. 6. What is XML?

Ans. XML stands for Extensible Markup Language and it is a markup language much like HTML. It is a meta language to let users design their own markup language and was designed to carry data, not to display data. XML tags are not predefined and You must define your own tags. XML is designed to be self-descriptive.

REVIEW QUESTIONS

1. What is HTML ? What do we need to create HTML?
2. What are the advantages and disadvantages of HTML?
3. Write a HTML Script to print "My India" on the screen.
4. Write a HTML Script to print a ordered list of 5 fruits on screen.
5. What is frame in HTML and how it is useful?
6. What is XML and how it is different form HTML?
7. How can be used XML in the aspects of web development?
8. How can be create a website using HTML?

❑❑❑

6

CHAPTER

Web Site Designing

INSIDE THIS CHAPTER

6.1. MAKING TO YOUR OWN WEBSITE

We can divide website making in several steps as follows:

1. Get Your Domain Name

The first thing you need to do before anything else is to get yourself a domain name. A domain name is the name you want to give to your website. For example, the domain name of the website you're reading is "internetfundamental.com". To get a domain name, you have to pay an annual fee to a registrar for the right to use that name. Getting a name does not get you a website or anything like that. It's just a name. It's sort of like registering a business name in the brick-and-mortar world; having that business name does not mean that you also have the shop premises to go with the name.

2. Choose a Web Host and Sign Up for an Account

A web host is basically a company that has many computers connected to the Internet. When you place your web pages on their computers, everyone in the world will be able to connect to it and view them. You will need to sign up for an account with a web host so that your website has a home. If getting a domain name is analogous to getting a business name in the brick-and-mortar world, getting a web hosting account is analogous to renting office premises for your business.

3. Designing your Web Pages

Once you have settled your domain name and web host, the next step is to design the web site itself. Although there are many considerations in web design, as a beginner, your first step is to actually get something out onto the web. The fine-tuning can come after you've figured out how to get a basic web page onto your site. One way is to use a web editor to do it. Such editors allow you to design your site visually, without having to muck around with the technical details. They work just like a normal word processor.

4. Testing Your Website

You will need to test your web pages as you design them in the major browsers: Internet Explorer 8, Internet Explorer 7, Internet Explorer 6, the latest versions of Firefox, Opera and Safari. Since all these browsers are free anyway, it should not be any hardship to get them and install them. The trick however, is testing with more than one version of Internet Explorer since the later version will overwrite the earlier.

5. Collecting Credit Card Information, Making Money

If you are selling products or services, you will need some way to collect credit card information.

6. Getting Your Site Noticed

When your site is ready, you will need to submit your site to the search engines, particularly Google. You can find the Google submission page by clicking on the "About Google" link on Google, and then locating the "Submit your content to Google" link on the page that appears. This is where promoting your website is important. This involves many things, including the usual way people did things before the Internet: advertisements in the newspapers, word-of-mouth, etc. You might want to consider advertising on places on internet like yahoo.com.

6.1.1. How to Choose a Good Domain Name

Here is some information and tips to help you select the best domain name for you:

1. **Keep it short :** The shorter, the better. It's best to keep your domain name less than 15 characters if possible. A shorter name is easier for people to remember and it also reduces the likelihood that your visitor will mistype it or spell it wrong. For example, BuyIt.com has a lot less chance of being mistyped than isellthispleasebuyitfrommetoday.com.
2. **Keep it sweet :** The more memorable the name, the easier it is for people to remember. You can have the best site in the world, but if people can't remember your domain name it will cut down on your return visitors. It's hard enough to get traffic on the web without "losing" visitors that want to come back but can't find you.
3. **Keep it simple :** Make sure the domain name is easy for people to spell. Don't spell words unusually. Unless you have incredible branding, people will type in the common spelling of your domain and your competition will get visitors who were looking for you. For example, BlueRight.com is much better than BluRite.com
4. **Choose a domain name that reflects your business :** Don't just choose a random name. For example, if your sell horse supplies, you should choose a name that has the word "horse supplies" in it such as YourHorseSupplies.com. Or, your domain name should refer to horses in some way, like whoa.com. The former will help you in the search engines. The latter will be memorable.
5. **When in doubt, choose a .com :** If you are debating between several different domain names, go with the .com. It's what most people think of when they are typing in a domain name. It's true the best names are gone, but you can still find a good .com if you look. Thousands of .coms expire and are released back in the market each month so there is always a steady supply of new domains coming into the market.
6. **Special tip :** If you can afford it, you may also want to get your name followed by .com along with whatever domain name you decide to get. I think the way the internet is going it is smart to do and you will be very glad you have it later on.

6.1.2. Choose a Web Host

Once you have your domain name, you will need to choose a web host. It is very important that you choose a good web host. Good doesn't mean expensive. Good means reliable with great support.

When you choose a web host, **make sure they have excellent support**. If you have a problem with your web site you want help now, not in a day or two.

So, how can you tell how responsive the web host you are considering actually is? You test them. If they have a phone number (and they should) call them late at night or on weekends to see if they are really there.

You can also send an e-mail to their support e-mail address (not their sales e-mail) and ask them a basic question like "How many e-mail addresses am I allowed?" or "Is there a limit on the number of auto responders I'm allowed?" and see how long it takes them to get back to you. For a non-emergency email like that I think up to 12 hours is acceptable. If you send the e-mail on Saturday and they don't get back to you until Monday, that isn't good. If they take more than 12 hours and either don't have phone support or aren't answering, choose another web host.

6.1.3. Do's and Don'ts of Making a Website

Regardless of what software you use to build a website, there are some basic do's and don'ts you need to keep in mind. It's very easy to get carried away on your first site and do things that send visitors away which is most definitely not what you want to do.

When you build a website:

1. **Do Keep Your Web Site Simple :** Don't cram it full of clutter. Unless it's a how-to site like this one, the less text, the better.
2. **Do make it Eye Pleasing :** No crazy fonts or colors. The standard fonts for websites are Arial, Verdana and Times New Roman. Whatever font you decide on, use it throughout your site. This site is in Arial. Don't alternate fonts.
3. **Don't use a Flash Introduction :** Web designers love this and if you use a website builder they can be really fun to create, but most people click off as soon as they see it loading. The object is to get people to see your website, not drive them away.
4. **Don't put music or sound files on your site**. It startles people when they are surfing and suddenly hear music blaring out. Their immediate reaction is to back track immediately.
5. **Don't put a lot of slow-loading graphics on your page**. No matter how pretty the graphics are, people aren't going to stick around and wait for them to load. You can optimize your images for free at NetMechanic.com. What this means is that you load the image onto Net Mechanic's web page and they will condense the size of the image for you, then let you pick the image you like from the results. Usually you can reduce the size by 50% or more without there being a noticeable change in quality.
6. **Do make sure visitors to your site can tell at a glance what your site is about.** For example, if you sell widgets, make sure that's evident right up front. Most people will leave in the first two seconds if they don't think your site offers them what they are looking for. They aren't going to take the time to scroll around to see if you have what they need. Make it immediately clear to them what your site is about.

7. **Don't use pop-ups on your website.** Visitors find these really annoying. The less you annoy a visitor, the more likely he or she is to stay.

6.1.4. How to Upload your Website to your Web Host Account

Once you build your website, you will need to get it to your web host's server. This process is called "uploading" your files. Some places call it "publishing your website."

If you are using the build a website software from your web host, all of your website files will already be on your web host's servers so all you will need to do is click a button. The button is usually Publish or Submit.

If you are using Microsoft's Expressions Web or another stand alone website builder, you will usually just need to log in to your web hosting account and click the upload files button. Some web hosts call this FTP, others just say Upload Files. It all depends on your web host what it's called but they both are the same thing.

6.2. MICROSOFT FRONTPAGE EXPRESS

These days Web authors have two choices - they can either learn HTML or code it by hand, or they can use one of the many software packages out there which aim to create your Website for you. The majority of these software programs are WYSIWYG (What You See is What You Get) editors, so design is often as easy as dragging and dropping elements onto a page and then configuring them to your liking.

The only problem with these programs is that they tend to cost a lot of money. What can you do if you don't want to touch HTML but would like to create your own site? FrontPage Express, from Microsoft, could be your answer.

Definition: ***FP Express is a cut down version of Microsoft's Web authoring software which provides a user-friendly interface to create web sites.***

FrontPage 98, and is available free for download from the Internet Explorer site.

6.2.1. Main features

There are several features of Microsoft FrontPage Express as follows:

1. As with many HTML editors, FrontPage Express is only pseudo-WYSIWYG. This means that you can't position elements exactly where you want to. FrontPage Express can solve this problem as you are asked a series of basic questions about how you would like your page to appear and a custom page is then generated for you, which you can edit as you see fit.
2. FrontPage Express acts just like a more advanced word processor in this respect - you type in any text you want, then format the font, colour, size and alignment. You can also insert pictures, form elements, tables or almost anything else.
3. FrontPage Express supports a large number of HTML features, including inline and background images, background sounds, videos, marquees, lists, tables, and form elements.
4. You can insert script into your pages with the *Insert Script* command, but you have to type the script in yourself. You can also insert ActiveX controls, Java applets, plug-ins, or

Powerpoint animations into your pages, but similarly you have to know the location of the appropriate file and exactly how it should be configured.

5. One interesting feature is the inclusion of WebBot components. These make use of FrontPage extensions on the Web server and provide increased functionality to Web pages, such as a search facility, a time stamp, and the ability to save data to a file.

6.2.2. Advantages of FrontPage Express

There are a number of good points to FrontPage Express.

1. Firstly, the WebBot components provide an easy way to add sophisticated features to Web pages - you can use the time stamp to automatically maintain a "last updated" line, for instance, or the form WebBot to log form results to a text file for viewing later.
2. Secondly, there is a feature which allows you to view the HTML of your page, colour coded to make it easier for you. Text, elements, attributes and attribute values are all distinguished to make the code clearer for beginners to understand.
3. An excellent feature is the estimated download time which appears in the status bar. It tells you the number of seconds which your page will take to download at a speed of 28.8Kbps, and takes into account text, images, tables, form elements and everything else on the page. If you are looking to keep your page as fast as possible, this is a good indication of how well you're doing.
4. FrontPage Express has a special command, *Insert Markup*, which allows you to enter HTML code yourself, and this is then marked as new code and is not checked for errors. Of course, you will not be able to preview the effects of your tag in the program, but it is a useful way of future-proofing.
5. Finally, the wizards are a useful time-saving feature if you need to construct a quick feedback form or survey for your Web page. They are also a good introduction to beginners on how Web pages are constructed and what they contain.

6.2.3. Disadvantages of FrontPage Express

The disadvantages to FrontPage Express are as follows:

1. One of the biggest disadvantage is that it is simply too basic. Cascading Style Sheets are not supported, so you can't use advanced formatting and layout techniques. Even frames are out of the question, both normal and floating frames, so you may have to author more pages than you need to.
2. Another disadvantage, though this one also applies to the commercial version, is that the WebBot components only work on Web servers which have the FrontPage Extensions installed. Most Web servers have yet to do this because of possible security risks and the fact that the extensions need to be run under Windows NT rather than UNIX, the more popular OS for servers. This means that the majority of users will miss out on the added functionality that the WebBots provide.
3. The nature of HTML editors means that there are often surplus tags in the code, such as an abundance of paragraph tags, or pointless META tags. This can make the code difficult to read, but can also inflate the size of your HTML file, so if you're looking to make the most compact code then it would be better for you to do it yourself.

6.2.4. Inserting a Background

We can insert a background by following the some steps as follows :

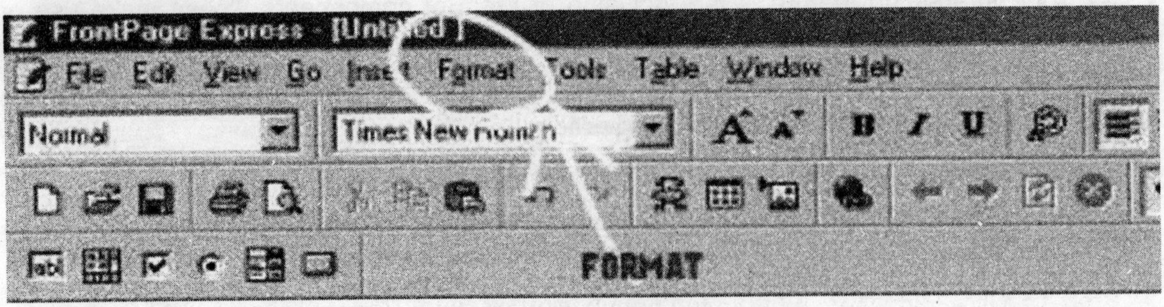

Tool Bar of FrontPage Express

Step 1 : On the uppermost tool bar click on format.

Step 2 : Next, scroll down and click on "background". A popup box will appear.

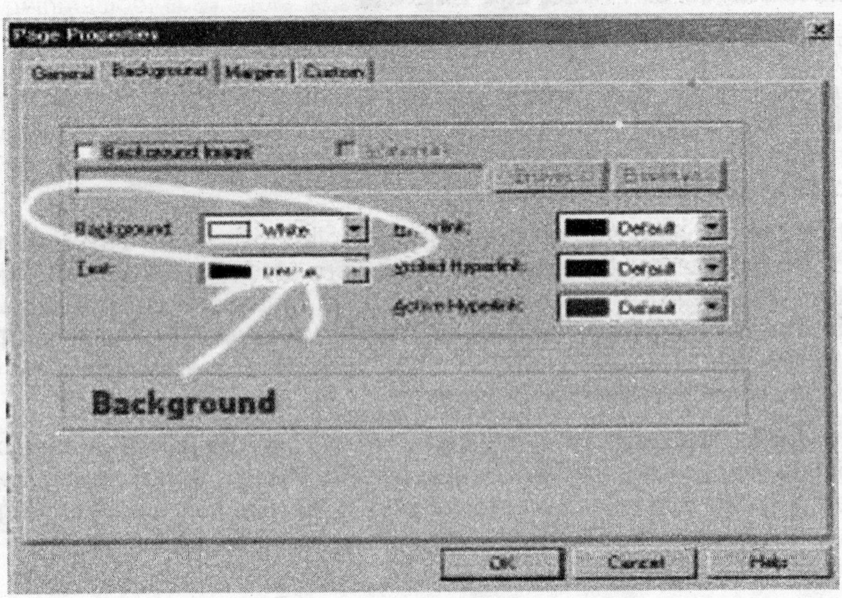

Step 3 : To the left of the first dropdown menu is background. Click on the dropdown menu and choose white from the color list.

Step 4 : Click ok and you now have a white background.

6.2.5. Inserting a Background Image

We can insert a background image by following the some steps as follows :

Step 1 : On the uppermost tool bar click on "Format.

Step 2 : Next, scroll down and click on "image". A popup box will appear.

Step 3 : This time check "background image."

Step 4 : Click "Browse" and from the popup window check "From File" and click "Browse" in this box.

Step 5 : Find the folder that your background image is in and click on the file.\

Step 6 : Click "Open".

Step 7 : You are now returned to the first popup window. "Click ok". Your background image is now on your page.

6.2.6. Inserting a Header Banner

Most WebPages open with some kind of banner at the top of the page. At the top I have my banner. You may choose to have an image that says Welcome or something else, but the steps are the same in all situations. We can insert a header banner by following the some steps as follows:

1. Click inside the page where you want your image to appear. For this tutorial click up ear the top of the page.
2. You can position the image either "left", "center" or "right", by clicking on the 3 boxes in the 2nd row of toolbars, to the right of the artists palette.(See toolbar image below. This image is "left".)

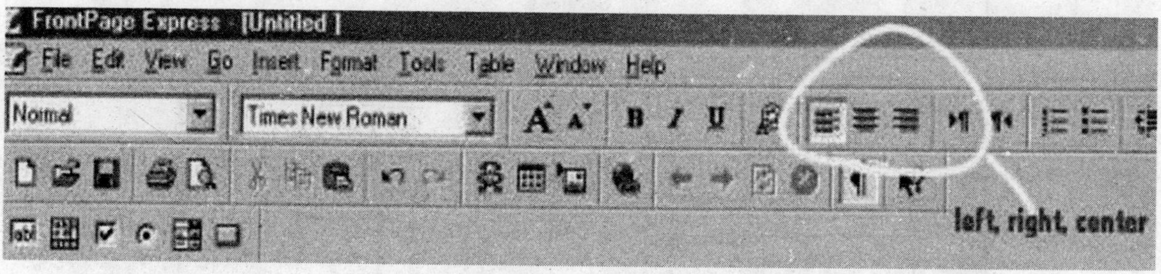

3. From the top toolbar, select "insert".
4. From the dropdown menu select "image".
5. From the popup window, check "file" and click "browse".
6. Find the folder that your background image is in and click on the file.
7. Click, "open" and click "ok". Image is now on page.

6.3. PLUG-INS [MDU May 2005, 2007, Dec. 2007]

Definition *Plug-in (also called plugin, addin, add-in, addon) are computer program that interacts with a host application (a web browser or an email client, for example) to provide a certain, usually very specific, function "on demand".*

Definition *Plug-ins are software modules that add to the functionality of an application. Many applications use plug-ins, including Eudora, Photoshop, and web browsers such as Firefox and Internet Explorer.*

How you install a plug-in depends on both the parent application and the plug-in itself. In many cases, the directory or folder where you place your plug-in(s) may have an installer. For more information, consult the documentation for both the plug-in and the application.

6.3.1. Need of Plug-Ins [MDU May 2005]

Plug-ins can have many different functions. Web browsers typically use plug-ins to expand the range of formats they can display (e.g., Acrobat documents, QuickTime or Adobe Flash content). When the web browser encounters a format it doesn't support, it looks for a plug-in that does. A plug-in can launch its own viewer, appear within a browser window, appear in an embedded frame (within a window that may also contain HTML), or run in the background. Plug-ins work in similar ways with other kinds of programs.

Applications support plug-ins for many reasons. Some of the main reasons include :

1. To enable third-party developers to create capabilities to extend an application
2. To support features yet unforeseen
3. To reduce the size of an application
4. To separate source code from an application because of incompatible software licenses.

6.3.2. Working of Plug-Ins [MDU May 2005]

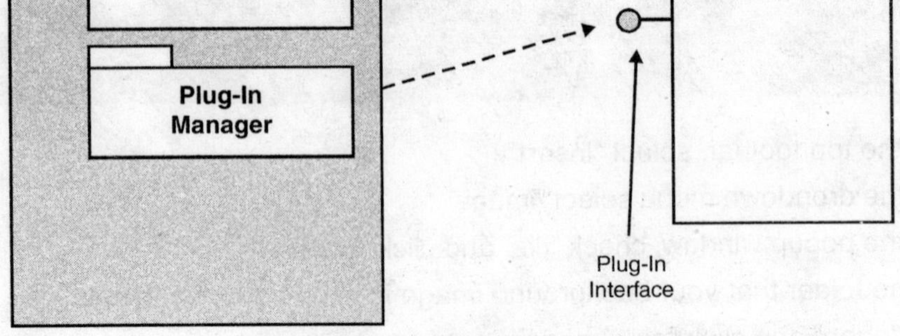

FIGURE 6.1. Working of Plug-Ins.

The host application provides services which the plug-in can use, including a way for plug-ins to register themselves with the host application and a protocol for the exchange of data with plug-ins. Plug-ins depend on the services provided by the host application and do not usually work by themselves. Conversely, the host application operates independently of the plug-ins, making it possible for end-users to add and update plug-ins dynamically without needing to make changes to the host application.

Open Application Programming Interfaces (APIs) provide a standard interface, allowing third parties to create plug-ins that interact with the host application. A stable API allows third-party plug-ins to continue to function as the original version changes and to extend the life-cycle of obsolete applications.

6.4. HELPER APPLICATIONS [MDU May 2005]

Definition

Helper Applications are actual stand-alone pieces of software that, when installed, will automatically start when needed by your web browser.

For example, let's say you are trying to view a movie that is being "streamed" over the Internet. If you clicked on a link for the movie on a web page and had the right Helper Application installed (say Real Player, for example), the browser would look for that software, start it, and play the video through it. Some Helper Applications work like plug-ins in that they can show the files in the same window the browser uses. However, a helper application can also run without the browser, and some of them can actually download files off the WWW and Internet without a web browser at all (but you have to know where those files are). Some of the most well-known helper applications are video ones: Real Player, Windows Media Player, and QuickTime Movie Player. Another well known helper application is Adobe Acrobat Reader which shows documents in versions that will remain the same no matter what computer they're on. (The federal and State Governments use Acrobat to publish forms on the WWW).

6.4.1. Plug-Ins vs. Helper Applications

A helper application is a separate application program that is invoked by the browser. It is simply a program that can understand and interpret files which the browser cannot handle by itself. Almost any program can be configured to act as a helper application for the browser. **Examples of helper applications include Telnet and Excel.**

The browser forks a separate process which starts the helper application. The helper application runs outside of the browser window. So, an advantage of helper applications over plug-ins is multitasking between a helper application and the browser window. If the browser is closed down, the helper application lives on.

Helper applications cannot display the contents of a file in the context of a Web page. If the file being read is a graphic, the helper application displays only the image, not the image embedded in the Web page. Another difference is that the browser has no control over the behavior of the helper application. The browser only has the ability to start the helper application and display the appropriate file.

A plug-in is an application program invoked by the browser. It is a dynamic code module designed to extend the capabilities of the browser by integrating a third party application program into the browser. Thus, a plug-in is part of the browser binary tree and runs inside the browser window. It cannot live on its own.

When plug-ins are installed they automatically tell the browser what file extensions they work with. Normally, there is no configuration involved with plug-ins, only installation. Because plug-ins are part of the browser binary tree they are platform specific. Therefore the correct version must be downloaded for plug-ins to work properly. Examples of plug-ins include RealAudio and Shockwave.

The integration of plug-ins into the browser is transparent to the user. Plug-ins simply open up and become active whenever the browser needs them. When the browser starts up, it checks for installed components. Perhaps you have seen the message "Loading plug-ins" when the browser starts up. If the browser receives a Web page that requires the use of a plug-in, the browser loads that plug-in. From that point on the browser should behave as if the plug-in is part of the browser itself. When you leave that particular Web page, the browser will discard the plug-in and free up all the memory it used.

So a helper application has a mind of its own, and a plug-in is literally plugged into the browser.

SUMMARY

1. Website designing is very creative job and required lot of creativity. Nowadays various tools are available to design and making of website.
2. The Microsoft FrontPage Express is one of them. Students will learn basics of web site designing and hosting by going through this chapter.
3. FP Express is a cut down version of Microsoft's Web authoring software which provides a user-friendly interface to create websites.
4. Plug-ins are software modules that add to the functionality of an application. Many applications use plug-ins, including Eudora, Photoshop, and web browsers such as Firefox and Internet Explorer.
5. When your site is ready, you will need to submit your site to the search engines, particularly Google.

SHORT QUESTION ANSWERS

Q. 1. How can you get your domain name?

Ans. The first thing you need to do before anything else is to get yourself a domain name. A domain name is the name you want to give to your website. For example, the domain name of the website you're reading is "internetfundamental.com". To get a domain name, you have to pay an annual fee to a registrar for the right to use that name. Getting a name does not get you a website or anything like that. It's just a name. It's sort of like registering a business name in the brick-and-mortar world; having that business name does not mean that you also have the shop premises to go with the name.

Q. 2. How can we ensure that our website getting noticed?

Ans. When your site is ready, you will need to submit your site to the search engines, particularly Google. You can find the Google submission page by clicking on the "About Google" link on Google, and then locating the "Submit your content to Google" link on the page that appears. This is where promoting your website is important. This involves many things, including the usual way people did things before the Internet: advertisements in the newspapers, word-of-mouth, etc. You might want to consider advertising on places on internet like yahoo.com.

Q. 3. How to Upload your Website to your Web Host Account?

Ans. Once you build your website, you will need to get it to your web host's server. This process is called "uploading" your files. Some places call it "publishing your website." Either way it's the same process.

If you are using the build a website software from your web host, all of your website files will already be on your web host's servers so all you will need to do is click a button. The button is usually Publish or Submit.

If you are using Microsoft's Expressions Web or another stand alone website builder, you will usually just need to log in to your web hosting account and click the upload files button. Some web hosts call this FTP, others just say Upload Files. It all depends on your web host what it's called but they both are the same thing.

Q. 4. What is Microsoft FrontPage Express?

Ans. FP Express is a cut down version of Microsoft's Web authoring software which provides a user-friendly interface to create web sites.

Q. 5. What do you understand by Plug-Ins?

Ans. Plug-ins are software modules that add to the functionality of an application. Many applications use plug-ins, including Eudora, Photoshop, and web browsers such as Firefox and Internet Explorer.

REVIEW QUESTIONS

1. How can you design your own website ?
2. How can you choose a good domain name ?
3. What are you supposed to do to host your website ?
4. Write Do's and Don'ts of making a website.
5. What is microsoft frontpage express ?
6. What are the advantages of frontpage express ?
7. Write main features of front page express ?
8. Write disadvantages of front page expression.

❑❑❑

LANGUAGES

SECTION- C

- Java Script
- Servers

SYLLABUS

Basic and Advanced HTML, Java Script Language, Client and Server Side Programming in Java Script. Forms and Data in Java Script, XML Basics. PWS, IIS, Apache; Microsoft Personal Web Server. Accessing and Using These Servers.

7

CHAPTER

Java Script

7.1. INTRODUCTION JAVASCRIPT

[MDU Dec. 2006]

INSIDE THIS CHAPTER

When the World Wide Web was first created in the early 1990s all web pages were static. When you viewed a web page you saw exactly what the page was set up to show you and there was no way for you to interact with the page.

Being able to interact with a web page - have it do something in response to your actions - required the addition of some form of programming language to "instruct" the page how it should respond to your actions. In order to have it respond immediately without having to reload the web page this language needed to be able to run on the same computer as the browser displaying the page.

At the time there were two browsers that were reasonably popular - Netscape Navigator and Internet Explorer. Netscape was the first to bring out a programming language that would allow web pages to become interactive - they called it **LiveScript** and it was integrated into the browser (meaning that the browser would interpret the commands directly without requiring the code to be compiled and without requiring a plugin to be able to run it). This meant that anyone using the latest Netscape browser would be able to interact with pages that made use of this language.

Another programming language called Java (which required a separate plugin in order to run) became very well known and so Netscape decided to try to cash in on this by renaming the language built into their browser to **JavaScript**. Note that while some Java and JavaScript code may appear similar, they are in fact two entirely different languages that serve completely different purposes.

Definition

JavaScript is a simple scripting language invented specifically for use in web browsers to make websites more dynamic.

Definition

JavaScript is an easy-to-use programming language that can be embedded in the header of your web pages. It can enhance the dynamics and interactive features of your page by allowing you to perform calculations, check forms, write interactive games, add special effects, customize graphics selections, and create security passwords and more.

JavaScript is most commonly used as a client side scripting language. This means that JavaScript code is written into an HTML page. When a user requests an HTML page with JavaScript in it, the script is sent to the browser and it's up to the browser to do something with it.

7.1.1. JavaScript vs. Java

JavaScript is *not* the same as Java. Although the names are much alike, JavaScript is primarily a scripting language for use within HTML pages, while Java is a real programming language that does quite different things from JavaScript. In addition Java is much harder to learn. It was developed by Sun for use in pretty much anything that needs some computing power. JavaScript can often be confused with Java due to the similar name. However, these are two different technologies .Java is an entire programming language developed by Sun Microsystems, while JavaScript is a scripting language that was introduced by Netscape. The difference can be better understood by following table :

JavaScript	Java
Interpreted (not compiled) by client. Compiled on server before execution on client.	Interpreted (not compiled) by client. Compiled on server before execution on client.
Object-based. Code uses built-in, extensible objects, but no classes or inheritance.	Object-oriented. Applets consist of object classes with inheritance.
Code integrated with, and embedded in, HTML.	Applets distinct from HTML (accessed from HTML pages).
Variable data types not declared (loose typing).	Variable data types must be declared (strong typing).
Dynamic binding. Object references checked at run-time.	Static binding. Object references must exist at compile-time.
Cannot automatically write to hard disk.	Cannot automatically write to hard disk.
Object-based. Code uses built-in, extensible objects, but no classes or inheritance.	Object-oriented. Applets consist of object classes with inheritance.
Code integrated with, and embedded in, HTML.	Applets distinct from HTML (accessed from HTML pages).

7.1.2. Advantages & Disadvantages of JavaScript

Like all computer languages, JavaScript has certain advantages and disadvantages. Many of the pros and cons are related to being a client-side language.

Advantages of JavaScript

- **Speed :** Being client-side, JavaScript is very fast because any code functions can be run immediately instead of having to contact the server and wait for an answer.
- **Simplicity :** JavaScript is relatively simple to learn and implement.
- **Versatility :** JavaScript plays nicely with other languages and can be used in a huge variety of applications. Unlike PHP or SSI scripts, JavaScript can be inserted into any web page regardless of the file extension. JavaScript can also be used inside scripts written in other languages such as Perl and PHP.
- **Server Load :** Being client-side reduces the demand on the website server.

Disadvantages of JavaScript

- **Security :** Because the code executes on the users' computer, in some cases it can be exploited for malicious purposes. This is one reason some people choose to disable JavaScript.
- **Reliance on End User :** JavaScript is sometimes interpreted differently by different browsers. Whereas server-side scripts will always produce the same output, client-side scripts can be a little unpredictable. Don't be overly concerned by this though - as long as you test your script in all the major browsers you should be safe.

7.2. EMBEDDING JAVASCRIPT INTO YOUR HTML DOCUMENT

Browsers that recognize JavaScript also recognize the special <SCRIPT> ... </SCRIPT> tag. This tag goes in the <HEAD> of your HTML document, along with your <TITLE> tag. Here's a short example:

```
<HTML>

<HEAD>

<TITLE>My JavaScript Page</TITLE>

<SCRIPT>
(JavaScript code goes in here)


</SCRIPT>

</HEAD>

(rest of your HTML document goes here)
```

To make an object (such as an image, a form button, or a hyperlink) on your page do something in response to a user action, you can add an additional attribute to the tag for that object. For example, the following HTML snippet pops up a thank you message when the user clicks on the link:

```
<A HREF="http://www.cs.brown.edu/courses/bridge/"
onClick="alert('Thanks for visiting the Bridge home
page!')">Bridge home page</A>
```

HREF="" attribute that specifies the URL that the link points to, but note the additional onClick="" attribute. The stuff between those quotation marks is JavaScript code, which in this case pops up an alert box with the specified message. (Don't worry if you don't understand the stuff between the quotation marks; you'll learn it later on. The important thing right now is to understand the additional attribute.)

Another point to recognize is that if a browser does not understand the <SCRIPT> tag, it will skip over it, and the text that is between the two tags (your code, basically) will appear on the screen as ordinary text. Since you want to create an HTML page that is viewable on all browsers, you would want to prevent this from happening.

To prevent this from happening, you should include the characters <!— immediately following the <SCRIPT> tag and the characters —> immediately preceding the </SCRIPT> tag. In HTML in general, using this set of characters allows you to write comments to yourself that the browser will not read or display on the page. Inside the JavaScript tags, the browser ignores these characters. By doing this, a non-JavaScript readable browser will see the comment notation and ignore the text between them (which is your code). Your code would then look like this:

```
<HEAD>
<SCRIPT>
<!—

(Your code goes here...)

—>
</SCRIPT>
</HEAD>
```

7.3. HOW TO WRITE JAVASCRIPT

Here are three important steps you should always follow when creating or using someone else's JavaScript code:

1. Use the script tag to tell the browser you are using JavaScript.
2. Write or download some JavaScript
3. Test the script.

There are so many different things that can go wrong with a script, be it human error, browser compatibility issues, or operating system differences. So, when using JavaScript, be sure that you test your script out on a wide variety of systems and most importantly, on different web browsers.

7.3.1. Your First JavaScript Script

To follow the classic examples of many programming languages, let's use JavaScript to print out "Hello World" to the browser. I know this isn't very interesting, but it will be a good way to explain all the overhead required to do something in JavaScript.

HTML and JavaScript Code:

```
<html>
<body>
<script type="text/JavaScript">
<!–
document.write("Hello World!")
//–>
</script>
</body>
</html>
```

Display:

```
Hello World!
```

Our first step was to tell the browser we were using a script with the <script> tag. Next we set the type of script equal to "text/JavaScript".

Next, we added an optional HTML comment that surrounds our JavaScript code. If a browser does not support JavaScript, it will not display our code in plain text to the user! The comment was ended with a "//—>" because "//" signifies a comment in JavaScript. We add that to prevent a browser from reading the end of the HTML comment in as a piece of JavaScript code.

JavaScript document.write

The final step of our script was to use a function called *document.write*, which writes a string into our HTML document. *document.write* can be used to write text, HTML, or a little of both. We passed the famous string of text to the function to spell out "Hello World!" which it printed to the screen.

Do not worry if you don't completely understand how *document.write* works, as we will be discussing functions in a later lesson.

TIPS

Looking at our JavaScript code above, notice that there is no semicolon at the end of the statement "document.write(Hello World!)". Why? JavaScript does not require that you use semicolons to signify the end of each statement.

TIPS

If you are an experienced programmer and prefer to use semicolons, feel free to do so. JavaScript will not malfunction from ending semicolons. The only time it is necessary to use a semicolon is when you choose to smash two statements onto one line(i.e. two* document.write *statements on one line).

7.4. VARIABLES

Definition

Variables are like storage units. You can create variables to hold values. It is ideal to name a variable something that is logical, so that you'll remember what you are using it for.

For example, if you were writing a program to divide 2 numbers, it could be confusing if you called your variables numberOne, numberTwo, numberThree because you may forget which one is the divisor, which one is the dividend, and which one is the quotient. A more logical approach would be to name them just that: divisor, dividend, quotient.

It is important to know the proper syntax to which variables must conform:

- They must start with a letter or underscore ("_")
- Subsequent characters can also be digits (0-9) or letters (A-Z and/or a-z). Remember, JavaScript is case-sensitive. (That means that MyVariable and myVariable are two different names to JavaScript, because they have different capitalization.)

Some examples of legal names are Number_hits, temp99, and _name.

When you declare a variable by assignment outside of a function, it is called a global variable, because it is available everywhere in the current document. When you declare a variable within a function, it is called a local variable, because it is available only within the function. Using var is optional, but you need to use it if you have a variable that has been declared global and you want to re-declare it as a local variable inside a function.

Variables can store all kinds of data (see below, Values of Variables, section 3.2). To assign a value to a variable, you use the following notation:

```
dividend = 8;

divisor = 4;

myString = "I may want to use this message multiple times";
message = myString;
```

Let's say the main part of the function will be dividing the dividend by the divisor and storing that number in a variable called quotient. I can write this line of code in my program: quotient = divisor*dividend, and I have both stored the value of the quotient to the variable quotient and I have declared the variable at the same time. If I had wanted to, I could have declared it along with my other assigned variables above, with a value of null. After executing the program, the value of quotient will be 2.

It is important to think about the design of your program before you begin. You should create the appropriate variables so that it makes your life easier when you go to write the program. For instance, if you know that you will be coding a lot of the same strings in a message, you may want to create a variable called message and give it the value of your message That way, when you call it in your program, you do not have to retype the same sentence over and over again, and if you want to change the content of that message, you only have to change it once — in the variable declaration.

7.4.1. Values of Variables

JavaScript recognizes the following types of values:

- Numbers, such as 42 or 3.14159
- Logical (Boolean) values, either true or false
- Strings, such as "Howdy!"
- null, a special keyword which refers to nothing

This relatively small set of types of values, or data types, enables you to perform useful functions with your applications. There is no explicit distinction between integer and real-valued numbers.

7.4.2. Data Type Conversion

JavaScript is a loosely typed language. That means you do not have to specify the data type of a variable when you declare it, and data types are converted automatically as needed during script execution. So, for example, you could define a variable as follows:

```
var answer = 42
```

And later, you could assign the same variable a string value, for example,

```
answer = "Thanks for all the fish..."
```

Because JavaScript is loosely typed, this assignment does not cause an error message. However, this is not good coding! You should create variables for a specific type, such as an integer, string, or array, and be consistent in the values that you store in the variable. This prevents confusion when you are writing your program.

In expressions involving numeric and string values, JavaScript converts the numeric values to strings. For example, consider the following statements:

```
x = "The answer is " + 42
y = 42 + " is the answer."
```

(The + sign tells JavaScript to concatenate, or stick together, the two strings. For example, if you write:

```
message = "Hello" + "World"
...then the variable message becomes the string "Hello World")
```

In the first statement, x becomes the string "The answer is 42." In the second, y becomes the string "42 is the answer."

7.5. LITERALS

You use literals to represent values in JavaScript. These are fixed values, not variables, that you literally provide in your script. Examples of literals include: 1234, "This is a literal," and true.

7.5.1. Integers

Integers can be expressed in decimal (base 10), hexadecimal (base 16), and octal (base 8). A decimal integer literal consists of a sequence of digits without a leading 0 (zero). A leading 0 (zero) on an integer literal indicates it is in octal; a leading 0x (or 0X) indicates hexadecimal. Hexadecimal integers can include digits (0-9) and the letters a-f and A-F. Octal integers can include only the digits 0-7.

Some examples of integer literals are: 42, 0xFFF, and -345.

7.5.2. Floating-point literals

A floating-point literal can have the following parts: a decimal integer, a decimal point ("."), a fraction (another decimal number), an exponent, and a type suffix. The exponent part is an "e" or "E" followed by an integer, which can be signed (preceded by "+" or "-"). A floating-point literal must have at least one digit, plus either a decimal point or "e" (or "E").

Some examples of floating-point literals are 3.1415, -3.1E12, .1e12, and 2E-12

7.5.3. Boolean literals

The Boolean type has two literal values: true and false.

7.5.4. String literals

A string literal is zero or more characters enclosed in double (") or single (') quotation marks. A string must be delimited by quotation marks of the same type; that is, either both single quotation marks or double quotation marks. The following are examples of string literals :

- "blah"
- 'blah'
- "1234"
- "one line \n another line"

In addition to ordinary characters, you can also include special characters in strings, as shown in the last element in the preceding list. The following table lists the special characters that you can use in JavaScript strings.

Character	Meaning
\b	backspace
\f	form feed
\n	new line
\r	carriage return
\t	tab
\\	backslash character

7.6. ARRAYS

Definition

An Array is an object which stores multiple values and has various properties. When you declare an array, you must declare the name of it, and then how many values it will need to store. It is important to realize that each value is stored in one of the elements of the array, and these elements start at 0.

First value in the array is really in the 0 element, and the second number is really in the first element. So for example, if I want to store 10 values in my array, the storage elements would range from 0-9. The notation for declaring an array looks like this:

```
myArray = new Array(10); foo = new Array(5);
```

Initially, all values are set to null. The notation for assigning values to each unit within the array looks like this:

```
myArray[0] = 56;
myArray[1] = 23;
myArray[9] = 44;
```

By putting the element number in brackets [ ] after the array's name, you can assign a value to that specific element. Note that there is no such element, in this example, as myArray[10]. Remember, the elements begin at myArray[0] and go up to myArray[9].

In JavaScript, however, an array's length increases if you assign a value to an element higher than the current length of the array. The following code creates an array of length zero, then assigns a value to element 99. This changes the length of the array to 100.

```
colors = new Array();
colors[99] = "midnightblue";
```

Because arrays are objects, they have certain properties that are pre-defined for your convenience. For example, you can find out how many elements myArray has and store this value in a variable called numberOfElements by using:

```
numberOfElements = myArray.length;
```

7.7. OPERATORS

JavaScript has many different operators, which come in several flavors, including binary. This tutorial will cover some of the most essential assignment, comparison, arithmetic and logical operators.

7.7.1. Selected Assignment Operators

An assignment operator assigns a value to its left operand based on the value of its right operand. The basic assignment operator is equal (=), which assigns the value of its right operand to its left operand. The other operators are shorthand for standard operations. Find an abridged list of shorthand operators below:

Shorthand operator	**Meaning**
x += y	x = x + y
x -= y	x = x - y
x *= y	x = x * y
x /= y	x = x / y

Note that the = sign here refers to assignment, not "equals" in the mathematical sense. So if x is 5 and y is 7, x = x + y is not a valid mathematical expression, but it works in JavaScript. It makes x the value of x + y (12 in this case).

EXAMPLES : If x = 10 and y = 2, x += y would mean x = x + y, hence x's new value would be the sum of x's previous value plus y's previous value. Upon executing x = x + y = 10 + 2, x's new value becomes 12, while y's new value remains equal to its old value. Similarly, x -= y would change x's value to 8. Calculate x *= y and x /= y to get a better idea of how these operators work.

7.7.2. Comparison Operators

A comparison operator compares its operands and returns a logical value based on whether the comparison is true or not. The operands can be numerical or string values. When used on string values, the comparisons are based on the standard lexicographical ordering. They are described in the following table.

Operator	Description	Example
Equal (= =)	Evaluates to true if the operands are equal.	x == y evaluates to true if x equals y.
Not equal (!=)	Evaluates to true if the operands are not equal.	x != y evaluates to true if x is not equal to y.
Greater than (>)	Evaluates to true if left operand is greater than right operand.	x > y evaluates to true if x is greater than y.
Greater than or equal (>=)	Evaluates to true if left operand is greater than or equal to right operand.	x >= y evaluates to true if x is greater than or equal to y.
Less than (<)	Evaluates to true if left operand is less than right operand.	x < y evaluates to true if x is less than y.
Less than or equal (<=)	Evaluates to true if left operand is less than or equal to right operand.	x <= y evaluates to true if x is less than or equal to y.

For Example 5 == 5 would return TRUE. 5 != 5 would return FALSE. (The statement 'Five is not equal to five.' is patently false.) 5 <= 5 would return TRUE. (Five is less than or equal to five. More precisely, it's exactly equal to five, but JavaScript could care less about boring details like that.)

7.7.3. Selected Arithmetic Operators

Arithmetic operators take numerical values (either literals or variables) as their operands and return a single numerical value. The standard arithmetic operators are addition (+), subtraction (-), multiplication (*), division (/) and remainder (%). These operators work as they do in other programming languages, as well as in standard algebra.

Since programmers frequently need to add or subtract 1 from a variable, JavaScript has shortcuts for doing this. myVar++ adds one to the value of myVar, while myVar— subtracts one from myVar.

For Example Let x = 3. x++ bumps x up to 4, while x— makes x equal to 2.

7.7.4. Logical Operators

Logical operators take Boolean (logical) values as operands and return a Boolean value. That is, they evaluate whether each subexpression within a Boolean expression is true or false, and then execute the operation on the respective truth values. Consider the following table:

Operator	Usage	Description
and (&&)	expr1 && expr2	True if both logical expressions expr1 and expr2 are true. False otherwise.
or (||)	expr1 || expr2	True if either logical expression expr1 or expr2 is true. False if both expr1 and expr2 are false.
not (!)	!expr	False if expr is true; true if expr is false.

Since we have now learned to use the essential operators, we can use them in conjunction with one another. See if you can work out why the following examples resolve the way they do:

If x = 4 and y = 7, ((x + y + 2) == 13) && (((x + y) / 2) == 2) returns FALSE.

If x = 4 and y = 7, ((y - x + 9) == 12) || ((x * y) == 2) returns TRUE.

If x = 4 and y = 7, !((x/2 + y) == 9) || ((x * (y/2)) == 2) returns FALSE.

7.8. USING JAVASCRIPT OBJECTS

When you load a document in your web browser, it creates a number of JavaScript objects with properties and capabilities based on the HTML in the document and other pertinent information. These objects exist in a hierarchy that reflects the structure of the HTML page itself.

The pre-defined objects that are most commonly used are the window and document objects. The window has methods that allow you to create new windows with the open() and close() methods. It also allows you to create message boxes using alert(), confirm(), and prompt(). Each displays the text that you put between the parentheses.

The document object models the HTML page. The document object contains arrays which store all the components constituting the contents of your web page, such as images, links, and forms. You can access and call methods on these elements of your web page through the arrays.

The objects in this pre-defined hierarchy can be accessed and modified. To refer to specific properties, you must specify the property name and all its ancestors, spelling out the complete hierarchy until the document object. A period, '.', is used in between each object and the name of its property. Generally, a property / object gets its name from the NAME attribute of the HTML tag. For example, the following refers to the *value* property of a text field named *text1* in a form named *myform* in the current document.

```
document.myform.text1.value
```

Form elements can also be accessed through the aforementioned forms array of the document object. In the above example, if the form named *myform* was the first form on the page, and *text1* was the third field in the form, the following also refers to that field's value property.

```
document.forms[0].elements[2].value
```

Functions (capabiltiies) of an object can similarly be accessed using the period notation. For example, the following instruction resets the 2nd form in the document.

```
document.forms[2].reset();
```

Click on one of the objects below to view the Netscape documentation on the specific properties and methods that that object has :

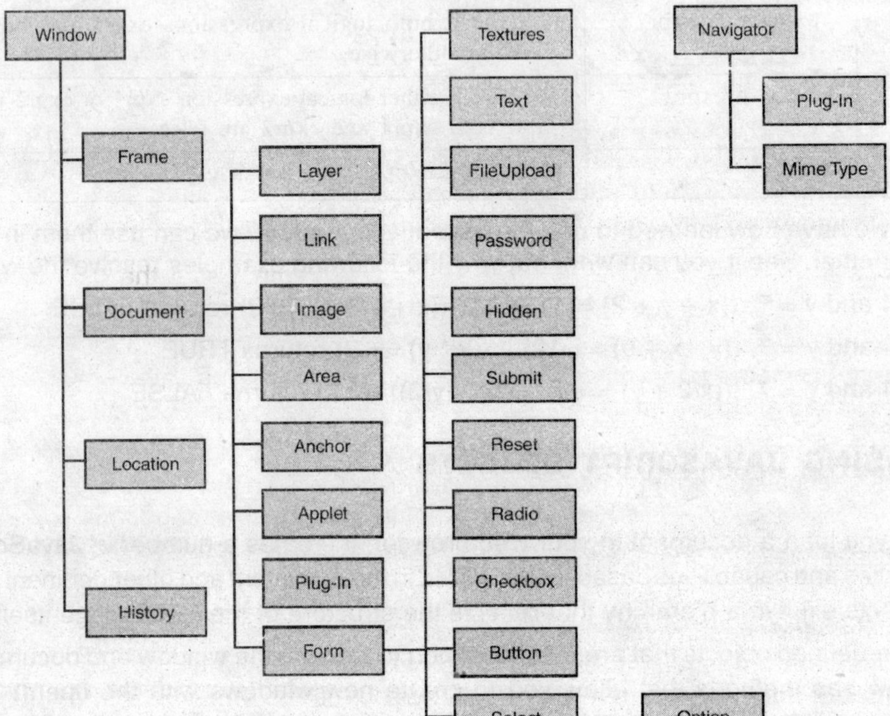

FIGURE 7.1. **JavaScript Objects.**

7.9. FUNCTIONS

Definition

Functions are one of the fundamental building blocks in JavaScript. A function is a JavaScript procedure — a set of statements that performs a specific task. A function definition has these basic parts:

- **The function keyword**
- **A function name**
- **A comma-separated list of arguments to the function in parentheses**
- **The statements in the function in curly braces: { }**

7.9.1. Defining a Function

When defining a function, it is *very* important that you pay close attention to the syntax. Unlike HTML, JavaScript is case sensitive, and it is very important to remember to enclose a function within curly braces { }, separate parameters with commas, and use a semi-colon at the end of your line of code.

It's important to understand the difference between *defining* and *calling* a function.

Defining the function names the function and specifies what to do when the function is called. You define a function within the <SCRIPT>...</SCRIPT> tags within the <HEAD>...</HEAD> tags. In defining a function, you must also declare the variables which you will be calling in that function.

Here's an example of *defining* a function:

```
function popupalert() {

        alert('This is an alert box.');

                                }
```

Notice the parentheses after the function name. It is imperative that you include these parentheses, even if they are empty. If you want to pass a *parameter* into the function, you would include that parameter inside of the parentheses. A parameter is a bit of extra information that can be different each time the function is run. It is stored in a variable and can be accessed just like any other variable. Here's an example of a function that takes a parameter:

```
function anotherAlert(word) {

        alert(word + ' is the word that you clicked on');

                                            }
```

When you call this function, you need to pass a parameter (such as the word that the user clicked on) into the function. Then the function can use this information. You can pass in a different word as a parameter each time you call the function, and the alert box will change appropriately. You'll see how to pass a parameter a little later on.

You can pass in multiple parameters, by separating them with a comma. You would want to pass in a few parameters if you have more than one variable that you either want to change or use in your function. Here are two examples of passing in multiple parameters when you are defining the function :

```
function secondAlert(word,password)    {

      confirm(word + ' is the word that you clicked on. The
```

```
secret password is ' + password);

                                        }

function thirdAlert(word,password)      {

  confirm(word + ' is the word you clicked on. Please

     take note of the password, ' + password);

                                        }
```

You'll notice that the same parameters are passed into both of these functions. However, you can pass in whatever values you want to use (see this same example below in calling the function).

7.9.2. Calling a Function

Calling the function actually performs the specified actions. When you call a function, this is usually within the BODY of the HTML page, and you usually pass a parameter into the function. A parameter is a variable from outside of the defined function on which the function will act.

Here's an example of calling the same function:

```
popupalert();
```

For the other example, this is how you may call it:

```
<A HREF="#top" onClick="anotherAlert('top')">top</A>
```

This would bring you to the top of the page, and bring up an alert box that said: "top is the word you clicked on" Try it for yourself: top

Here is the same example with multiple parameters that was shown above:

```
<A HREF="#top"
onClick="secondAlert('awesome','pandas')">awesome</A>

<A HREF="#top"
onClick="thirdAlert('computers','insert')">computers</A>
```

You'll notice in the code that different values for the variables word and password are passed in. These values here are what the function will need to perform the actions in the function. Make sure that the values you pass in are in the correct order because the function will take them in and assign these values to the parameters in the parentheses of the function declaration. Once you pass values into your function, you can use them however you want within your function.

7.10. IF/ELSE STATEMENTS

if statements execute a set of commands if a specified condition is true. If the condition is false, another set of statements can be executed through the use of the else keyword.

As you can see, the idea is quite intuitive and, surprisingly enough, so is the syntax:

```
if (condition) {
    statements1
}
-or-
if (condition) {
    statements1
}
else {
    statements2
}
```

(An **if** statement does not require an **else** statement following it, but an **else** statement must be preceded by an **if** statement.)

condition can be any JavaScript expression that evaluates to true or false. Parentheses are required around the condition. If *condition* evaluates to true, the statements in *statements1* are executed.

statements1 and *statements2* can be any JavaScript statements, including further nested **if** statements. Multiple statements must be enclosed in braces.

Here's an example:

```
if (weather == 'good') {
   go_out(we);
   have_a_picnic(we);
   do_cartwheels(Lisa);
}
else {
   stay_in(we);
   watch_TV(Catherine);
}
```

7.11. LOOPS

Definition

Loops are an incredibly useful programming tool. Loops handle repetitive tasks extremely well, especially in the context of consecutive elements.

The two most common types of loops are for and while loops :

7.11.1. For Loops

A for loop constitutes a statement which consists of three expressions, enclosed in parentheses and separated by semicolons, followed by a block of statements executed in the loop.

This definition may, at first, sound confusing. Indeed, it is hard to understand for loops without seeing them in action.

A for loop resembles the following:

```
for (initial-expression; condition; increment-expression) {
    statements
}
```

The **initial-expression** is a statement or variable declaration. (See the section on variables for more information.) It is typically used to initialize a counter variable. This expression may optionally declare new variables with the var keyword.

The **condition** is evaluated on each pass through the loop. If this condition evaluates to true, the statements in *statements* are performed. When the condition evaluates to false, the execution of the for loop stops. This conditional test is optional. If omitted, the condition always evaluates to true.

The **increment-expression** is generally used to update or increment the counter variable.

The *statements* constitute a block of statements that are executed as long as *condition* evaluates to true. This can be a single statement or multiple statements. Although not required, it is good practice to indent these statements from the beginning of the for statement to make your code more readable.

Check out the following for statement. It starts by declaring the variable *i* and initializing it to zero. It checks whether *i* is less than nine, performs the two successive statements, and increments *i* by one after each pass through the loop:

```
var n = 0;
for (var i = 0; i < 3; i++) {
    n += i;
    alert("The value of n is now " + n);
}
```

Try it yourself: Click this link

7.11.2. While Loops

The while loop, although most people would not recognize it as such, is for's twin. The two can fill in for one another - using either one is only a matter of convenience or preference according to context. While creates a loop that evaluates an expression, and if it is true, executes a block of statements. The loop then repeats, as long as the specified condition is true.

The syntax of while differs slightly from that of for:

```
while (condition) {
    statements
}
```

condition is evaluated before each pass through the loop. If this condition evaluates to true, the statements in the succeeding block are performed. When *condition* evaluates to false, execution continues with the statement following *statements*.

statements is a block of statements that are executed as long as the *condition* evaluates to true. Although not required, it is good practice to indent these statements from the beginning of the statement.

The following while loop iterates as long as *n* is less than three.

```
var n = 0;
var x = 0;
while(n < 3) {
    n++;
    x += n;
    alert("The value of n is " + n + ". The value of x is " +
x);
}
Try it for yourself: Click this link
```

7.12. COMMENTING

Comments allow you to write notes to yourself within your program. These are important because they allow someone to browse your code and understand what the various functions do or what your variables represent. Comments also allow you to understand your code if it's been a while since you last looked at it.

In JavaScript, you can write both one-line comments and multiple line comments. The notation for each is different though. For a one line comment, you precede your comment with //. This indicates that everything written on that line, after the //, is a comment and the program should disregard it.

For a multiple-line comment, you start with /* and end with */ . It is nice to put an * at the beginning of each line just so someone perusing your code realizes that he/she is looking at a comment (if it is really long this helps). This is not necessary though.

The following are examples of comments in JavaScript.

```
// This is a single line comment.

/* This is a multiple line comment with only one line. */

/* This is a multiple line comment.
```

```
* The star (*) at the beginning of this line is optional.
* So is the star at the beginning of this line. */
```

7.13. JAVASCRIPT STRING BASICS

Definition ***The string in JavaScript - similar to other programming languages - is a variable which stores a series of characters.***

A string can be any text inside a quote pair. You can use simple or double quotes, but of course the pairs must match. The next code example shows how you can define strings in JavaScript:

```
var str1 = 'Test string 1';
var str2 = "Test string 2";
var str3 = 'It\'s a string';
var str4 = "This is an 'internal' string";
var str5 = 'This is also an "internal" string';
```

As you can see in case of str3 we used a single quote to define our string content, but the text itself also contains a single quote. In such cases you need to use a backslash character before the single quote to display your text properly.

7.13.1. Concatenating Strings

You can also create a string by concatenating more strings together. This could be very useful if you build up your string within an application logic. To do this you have 2 operators to use. The + and the +=. The example below shows you how to use them.

```
var str1 = 'This ' + 'is ' + 'a ' + 'demo!';
var str2 = 'Part 1';
str2 += ' and Part 2';

var str3 = 'Mixed ';
str3 += 'mode ' + 'is possible.';
```

7.13.2. Strings and Numbers

JavaScript is very relaxed about the difference between strings and numbers. Some programming languages require you to state if a variable is a number or a string before doing anything else with it. Not so in JavaScript. In fact you can even add up numbers and strings:

```
var c = a + 12;
```

Some programming languages would shut down in disgust when they encounter this line. After all, a is a string and 12 is a number. JavaScript, however, tries to solve the problem by assuming 12 is also a string. So c becomes

```
Hello world!12
```

So if you use + on a string and a number, JavaScript is going to make the number a string for you. Better still, if you need it you can treat numbers as strings or strings as numbers.

Conversely, if you apply mathematics to a string, JavaScript tries to make it a number. If the string cannot be interpreted as a number (because there are letters in it, for instance), JavaScript gives NaN (Not a Number).Finally, JavaScript makes no reliable difference between integers and floating point variables.

Number to String

ToString() does not work in Netscape 2 and Explorer 3.

To convert a number to a string, do:

```
var c = (16 * 24)/49 + 12;
d = c.toString();
```

Now you can use all methods of strings on d, while c still contains the number.

String to Number

If you want to change a string to a number, first make sure there are only the characters 0-9 in the string. What I always do is simply multiplying the string by 1.

```
var c = '1234';
d = c * 1;
```

Since multiplying assumes numbers, JavaScript makes the string a number, if possible. If that isn't possible, the result is NaN. Note that you cannot do

```
var c = '1234';
d = c + 0;
```

This would yield '12340' because JavaScript thinks the + means *concatenate string*, not *add*.

7.13.3. String methods and properties

So what can we do with strings? Here's a list of automatic JavaScript commands. The concatenation is a special case, but all other commands can be used on any string by doing stringname.command().

- Concatenating strings (+)
- indexOf
- lastIndexOf
- charAt
- length
- split
- substring
- substr
- toLowerCase and toUpperCase

1. Concatenating strings (+)

First of all, you can concatenate strings, sew them together as it were:

```
document.write(a + b);
```

gives

```
Hello world!I am a JavaScript hacker.
```

But of course you want a space between the two sentences. No problem. By doing

```
document.write(a + ' ' + b);
```

we concatenate three strings: a, ' ' (one space) and b and we get

```
Hello world! I am a JavaScript hacker.
```

You can even use numbers or calculations, like:

```
document.write (a + 3*3 + b);
```

Now we concatenate string a, then the result of 3*3, treated as a string for the occasion, and then b, which gives

```
Hello world!9I am a JavaScript hacker.
```

Watch out if you want to add numbers. This

```
document.write (a + 3+3 + b);
```

concatenates four strings: a, 3, 3 and b, because + in this context means *concatenate strings*, not *add*, and gives

```
Hello world!33I am a JavaScript hacker.
```

Use brackets if you want to add 3 and 3 before making the string.

```
document.write (a + (3+3) + b);
```

Concatenates a, the result of 3+3 and b and gives

```
Hello world!6I am a JavaScript hacker.
```

2. indexOf

One of the most widely used automatical methods is indexOf. Each character has an *index* number, giving its position in the string. Note that the first character has index 0, the second one index 1 etc. So the index of 'w' in string a is 6.

Using indexOf we can read out the index number of a character. Put the .indexOf('') behind the string name and put the character you're looking for between the quotes. This

```
var a = 'Hello world!';
document.write(a.indexOf('w'))
```

gives 6. If the same character occurs more than once, this method gives the first occurrence. So

```
document.write(a.indexOf('o'))
```

gives 4 because that's the index of the first 'o' in the string.

You can also look for a combination of characters (This is of course also a string, but to prevent confusion I won't call it so). indexOf returns the position of the first character of the combination. So

```
document.write(a.indexOf('o w'))
```

also gives 4 because that's the index of the 'o'.

Furthermore it's possible to search for a character only after a certain index number. For instance, if you'd do

```
document.write(a.indexOf('o',5));
```

you get the index number of the first 'o' coming after the character with index 5 (= the space), so it would give 7.

Finally, if the character or combination is not present in the string, indexOf gives -1. This is actually the widest use of indexOf: seeing if a certain combination of characters is present. It's the core of the browser detect script. To see if the browser is Explorer, you take the string navigator.userAgent and see if it contains 'MSIE':

```
if (navigator.userAgent.indexOf('MSIE') !=-1)
{
    do something for Explorer
}
```

If the index number of 'MSIE' is *not* -1 (if 'MSIE' occurs anywhere in the string), the browser is Explorer.

3. lastIndexOf

There's also lastIndexOf which gives the last occurrence of the character or combination. It otherwise works exactly as indexOf. This

```
var b = 'I am a JavaScript hacker.'
document.write(b.lastIndexOf('a'))
```

gives 19 because that's the index of the last 'a' in the string.

4. charAt

charAt() gives you the character at a certain position. For instance, when you do

```
var b = 'I am a JavaScript hacker.'
document.write(b.charAt(5))
```

gives 'a', because that's the character at the sixth position (remember, first character is 0!).

5. length

length gives you the length of a string.

```
var b = 'I am a JavaScript hacker.'
document.write(b.length)
```

gives 25. Note that the length is 1 more than the index number of the last character.

6. split

split() does not work in Netscape 2 and Explorer 3.

split() is a specialized method that you need sometimes. It allows you to split a string at the places of a certain character. You must put the result in an array, not in a simple variable. Let's split b on the spaces.

```
var b = 'I am a JavaScript hacker.'
var temp = new Array();
temp = b.split(' ');
```

Now the string has been split into 5 strings that are placed in the array temp. The spaces themselves are gone.

```
temp[0] = 'I';
temp[1] = 'am';
temp[2] = 'a';
temp[3] = 'JavaScript';
temp[4] = 'hacker.';
```

7. substring

substring is used to take a part of a string. Syntax is substring(first_index,last_index). So for instance

```
var a = 'Hello world!';
document.write(a.substring(4,8));
```

gives 'o wo', from the first 'o' (index 4) to the second one (index 7) Note that the 'r' (index 8) is *not* part of this substring.

You can also do

```
var a = 'Hello world!';
document.write(a.substring(4));
```

This gives the whole string from the character with index 4 on: 'o world!'.

8. substr

substr() is not supported by Netscape 2 and 3, Explorer 3, Hotjava 3, Opera 3 and WebTV.

There is also a method substr() that works slightly differently. Instead of the second number being an index number, it gives the number of characters. So

```
document.write(a.substr(4,8));
```

starts at the character with index 4 ('o') and then gives 8 characters, so the output is

```
o world!
```

Since substr() is not supported by the Version 3 browsers I never use it.

9. toLowerCase and toUpperCase

Finally two methods that you need sometimes: toLowerCase() puts the whole string in lower case, toUpperCase() puts the whole string in upper case.

```
var b = 'I am a JavaScript hacker.'
document.write(b.toUpperCase())
```

gives 'I AM A JAVASCRIPT HACKER.'.

7.14. CLIENT-SIDE SCRIPTING

Definition

Most web applications involve both client-side and server-side script. Client-side script is often used to program the user interface for an application - for example, to change a web page's text dynamically, respond to user actions such as button clicks, and perform such client-oriented tasks as input validation. Client-side script executes locally in the browser, which provides the user with a lively and responsive interface.

Server-side script, in contrast, is usually used to program an application's back end. This often involves accessing a database or performing middle-tier business logic. Server-side script is also used to create applications which need a broad reach: applications that might be accessed by many different types of browsers, each with different capabilities.

But client and server script are generally independent. When a page is first requested, the server can run server-side script, and then pass the page to the browser, which can then run client-side script. However, if a server-side script on the page must be run again, the page must be submitted back to the server, which effectively restarts the page. Maintaining the state of the controls and values on the page can involve complex scripting to pass information back and forth between the browser and server. In addition, the roundtrip between client and server involves overhead (most notably the upload and download time) that can slow an application.

The following are some techniques that can be used to help reduce some of this overhead:

1. Using Javascript Array
2. Using ASP and XML
3. Using Remote Scripting

7.14.1. Using JavaScript and forms

Definition

Forms are used on the web to collect information from the user.

This can be anything from a Google search box through to an online application for a financial product. They take hyperlinks to a higher level, by allowing different things to happen depending on what information the form captures. A simple example is Google - entering two different keywords into the search textfield produces two different lists of returned results.

With JavaScript at your side, you can process simple forms without invoking the server. And when submitting the form to a CGI program is necessary, you can have JavaScript take care of all the preliminary requirements, such as validating input to ensure that the user has dotted every i. In this column we'll look closely at the JavaScript-form connection, including how to use JavaScript's form object, how to read and set form content, and how to trigger JavaScript events by manipulating form controls. We'll also cover how to use JavaScript to verify form input and sumbit the form to a CGI program.

7.14.2. Creating the Form

There are few differences between a straight HTML form and a JavaScript-enhanced form. The main one being that a JavaScript form relies on one or more event handlers, such as onClick or onSubmit. These invoke a JavaScript action when the user does something in the form, like clicking a button. The event handlers, which are placed with the rest of the attributes in the HTML form tags, are invisible to a browser that doesn't support JavaScript. Because of this trait, you can often use one form for both JavaScript and non-JavaScript browsers.

Typical form control objects include the following:

- Text box for entering a line of text
- Push button for selecting an action
- Radio buttons for making one selection among a group of options
- Check boxes for selecting or deselecting a single, independent option

For use with JavaScript, you should always remember to provide a name for the form itself, and each control you use. The names allow you to reference the object in your JavaScript-enhanced page.

The typical form looks like this.

```
<FORM NAME="myform" ACTION="" METHOD="GET">
```

Enter something in the box:

```
<INPUT TYPE="text" NAME="inputbox" VALUE=""><P>
<INPUT TYPE="button" NAME="button" Value="Click"
onClick="testResults(this.form)">
</FORM>
```

- ***FORM NAME="myform"*** defines and names the form. Elsewhere in the JavaScript you can reference this form by the name *myform*. The name you give your form is up to you, but it should comply with JavaScript's standard variable/function naming rules (no spaces, no weird characters except the underscore, etc.).
- ***ACTION=""*** defines how you want the browser to handle the form when it is submitted to a CGI program running on the server. As this example is not designed to submit anything, the URL for the CGI program is omitted.
- ***METHOD="GET"*** defines the method data is passed to the server when the form is submitted. In this case the atttibute is puffer as the example form does not submit anything.
- ***INPUT TYPE="text"*** defines the text box object. This is standard HTML markup.

- ***INPUT TYPE="button"*** defines the button object. This is standard HTML markup except for the onClick handler.
- ***onClick="testResults(this.form)"*** is an event handler — it handles an event, in this case clicking the button. When the button is clicked, JavaScript executes the expression within the quotes. The expression says to call the testResults function elsewhere on the page, and pass to it the current form object.

7.14.3. Getting a Value From a Form Object

Here's the full script you can try as you experiment with obtaining values from form objects. Load the page, then type something into the text box. Click the button, and what you typed is shown in the alert box.

testform.html

```
<HTML>
<HEAD>
<TITLE>Test Input</TITLE>
<SCRIPT LANGUAGE="JavaScript">
function testResults (form) {
   var TestVar = form.inputbox.value;
   alert ("You typed: " + TestVar);
}
</SCRIPT>
</HEAD>
<BODY>
<FORM NAME="myform" ACTION="" METHOD="GET">Enter something in
the box: <BR>
<INPUT TYPE="text" NAME="inputbox" VALUE=""><P>
<INPUT TYPE="button" NAME="button" Value="Click"
onClick="testResults(this.form)">
</FORM>
</BODY>
</HTML>
```

Here's how the script works. JavaScript calls the testResults function when you click the button in the form. The testResults function is passed the form object using the syntax this.form (the this keyword references the button control; form is a property of the button control and represents the form object). I've given the form object the name *form* inside the testResult function, but you can any name you like.

The testResults function is simple — it merely copies the contents of the text box to a variable named TestVar. Notice how the text box contents was referenced. I defined the form object I wanted to use (called *form*), the object within the form that I wanted (called *inputbox*), and the property of that object I wanted (the *value* property).

7.14.4. Setting a Value in a Form Object

The value property of the inputbox, shown in the above example, is both readable and writable. That is, you can read whatever is typed into the box, and you can write data back into it. The process of setting the value in a form object is just the reverse of reading it. Here's a short example to demonstrate setting a value in a form text box. The process is similar to the previous example, except this time there are two buttons. Click the "Read" button and the script reads what you typed into the text box. Click the "Write" button and the script writes a particularly lurid phrase into the text box.

set_formval.html

```
<HTML>
<HEAD>
<TITLE>Test Input </TITLE>
<SCRIPT LANGUAGE="JavaScript">
function readText (form) {
   TestVar =form.inputbox.value;
   alert ("You typed: " + TestVar);
}

function writeText (form) {
   form.inputbox.value = "Have a nice day!"
}
</SCRIPT>
</HEAD>
<BODY>
<FORM NAME="myform" ACTION="" METHOD="GET">
Enter something in the box: <BR>
<INPUT TYPE="text" NAME="inputbox" VALUE=""><P>
<INPUT TYPE="button" NAME="button1" Value="Read"
onClick="readText(this.form)">
<INPUT TYPE="button" NAME="button2" Value="Write"
onClick="writeText(this.form)">
</FORM>
</BODY>
</HTML>
```

- When you click the "Read" button, JavaScript calls the readText function, which reads and displays the value you entered into the text box.

- When you click the "Write" button, JavaScript calls the writeText function, which writes "Have a nice day!" in the text box.

7.14.5. Reading Other Form Object Values

The text box is perhaps the most common form object you'll read (or write) using JavaScript. However, you can use JavaScript to read and write values from most other objects (note that JavaScript cannot currently be used to read or write data using the password text box). In addition to text boxes, JavaScript can be used with:

- Hidden text box (TYPE="hidden").
- Radio button (TYPE="radio")
- Check box (TYPE="checkbox")
- Text area (<TEXT AREA>)
- Lists (<SELECT>)

7.14.6. Other Events You Can Trigger Within a Form

I have used the onClick event handler in all of the examples because that's the one you are most likely to deal with in your forms. Yet JavaScript supports a number of other event handlers as well. Use these as the need arises, and the mood fits. In Navigator 2.x The event handlers used with form object are:

- *onFocus* — an event is triggered with a form object gets input focus (the insertion point is clicked there).
- *onBlur* — an event is triggered with a form object loses input focus (the insertion point is clicked out of there).
- *onChange* — an event is triggered when a new item is selected in a list box. This event is also trigged with a text or text area box loses focus and the contents of the box has changed.
- *onSelect* — an event is triggered when text in a text or text area box is selected.
- *onSubmit* — an event is triggered when the form is submitted to the server (more about this important hander later in the column).

7.14.7. Submitting the Form to the Server

In the examples above I've limited the action of the form to within JavaScript only. Many forms are designed to send data back to a CGI program runnong on the server. This is referred to as submitting the form, and is accomplished using either of two JavaScript instructions: the onSubmit event handler or the submit method. In most instances, you use one or the other, *not both*!

- Place the onSubmit event hander in the <FORM> tag. This tells JavaScript what it should do when the user clicks the Submit button (this is a button defined as TYPE="submit").
- Place the submit instruction anywhere in your JavaScript. It can be activated by any action, such as clicking a form button that has been defined with the onClick event handler.

7.14.8. Validating Form Data Using JavaScript

The World Wide Web "grew up" when they added the ability to display forms. In the days before forms, the Web was only mildly interactive, with just hypertext links to take readers from location to location. Forms allow users to truly interact with the Web. For example, readers can specify search queries using a simple one-line text box.

The typical form as used on a Web page consists of two parts: the form itself, which is rendered in the browser, and a CGI script or program located on the server. This script processes the user's input. While it's not exactly rocket science, a stumbling block in creating great Web forms is writing the CGI program. In most cases these programs are written in Perl or C, and can be a bother to implement and debug. A primary job of the CGI program is to validate that the reader has provided correct data, and this can requires pages of code.

JavaScript changes that. With JavaScript you can check the data provided by the reader before it's ever sent to the CGI program. In this way the CGI program can be kept to a bare minimum. And, because the data is only sent after it has been validated, the server need not be bothered until the form entry is known to be good. This saves valuable server resources.

Most form validation chores revolve around basic data checking: does the user remember to fill in an box? Is it have the right length? Does it contain valid characters? With most forms you can readily answer these questions with a small handful of validation routines.

A typical validation routine is determining if an input box contains only numeric digits, shown below. If the entry contains non-numeric characters, you can ask the user to enter the correct data. A readymade routine for this is the isNumberString function, which returns the value 1 if the string contains only numbers, and 0 if it contains any non-numeric characters. To use it, provide the data string as the parameter. The value returned by the function tells you if the data is valid.

valid_simple.html

```
<HTML>
<HEAD>
<TITLE>Test Input Validation</TITLE>
<SCRIPT LANGUAGE="JavaScript">
function testResults (form) {
    TestVar = isNumberString (form.inputbox.value)
     if (TestVar == 1)
        alert ("Congratulations! You entered only numbers");
     else
        alert ("Boo! You entered a string with non-numbers
characters");
}

function isNumberString (InString)  {
    if(InString.length==0) return (false);
```

```
    var RefString="1234567890";
    for (Count=0; Count < InString.length; Count++)  {
        TempChar= InString.substring (Count, Count+1);
        if (RefString.indexOf (TempChar, 0)==-1)
            return (false);
    }
    return (true);
}
</SCRIPT>
</HEAD>
<BODY>
<FORM NAME="myform">
Enter a string with numbers only:
<INPUT TYPE="text" NAME="inputbox" VALUE="">
<INPUT TYPE="button" NAME="button" Value="Click"
onClick="testResults(this.form)" >
</FORM>
</BODY>
</HTML>
```

7.15. INTRODUCTION SERVER SIDE SCRIPTING [MDU May 2008]

Definition

Server-Side Scripts are small programs that reside on the web server to interact with the web environment. When browsers execute scripts that link to the code file, the script will do its designed functions and return a dynamically generated page.

Typical use of server-side programming :

- Personalized page content to the user's needs
- Enable the web browser to act as a database interface
- E-Commerce shopping carts

Another advantage of server side scripts is that they can have access to files and databases that would not normally be available to a visitor (as well as the client side scripts). This makes them more flexible in the way they use and access various data sources, as well as being more secure in the sense that the data sources do not need to be web accessible and so exposed.

Finally, server side scripts can be written in a much wider variety of languages than their client side counterparts. PHP, Perl, VBScript, PerlScript, Python, C++, Java and JScript to name but a few. Of course, which language you choose to script in will depend on your server configuration.

Ideas for when to use server side scripts.

- Password protection.
- Browser sniffing/customization.
- Form processing.
- Building and displaying pages created from a database.

7.15.1. Difference in Client–Side and Server Scripting

With Client-Side Scripting, the scripts on the page are processed by the individual web browser that requests the page. **With Server-Side Scripting,** on the other hand, all scripts are processed on the server before the requested page is ever sent to the browser. This results in no actual code being sent to the client's machine.

One of the most popular methods used to create today's modern dynamic web pages is Server-Side Scripting languages. These dynamic pages are constructed in such a way that all server processes take place before the page is delivered to the user. This means that you only need the most basic internet browsing software to view the most complex and dynamic pages on the web today.

Contrary to Server-Side scripts, Client-Side scripts can only run on browsers that support the specific scripting language that you are using. This means that you would need Netscape Navigator to view fully functioning JavaScript. On the contrary, you would need Internet Explorer to view fully functioning VB Script.

Server-Side Scripting has made it possible to "create platform-independent, easily deployable applications". The thought of creating a program or application that will run anywhere in the world has obvious advantages over Client-Side Scripting.

7.16. JAVASCRIPT ON THE SERVER [MDU May 2006]

JavaScript extends the capabilities of the server. By providing a scripting language the software can do more without calling an external program. This makes it easier for Webmasters to add features to their sites that browsers can take advantage of. And it can reduce the load on the server by keeping the processing within the server software.

7.16.1. Request Object

Each time a browser wants more information it sends a request to a server. Data about these requests are available on the server from the request object in JavaScript. If you have written CGI scripts, you should be familiar with the properties that are built into the request object.

7.16.1.1. The Built-In Properties

The request object is initialized with four properties: agent, ip, method, and protocol. A JavaScript function on the server can use this information to determine its response to the request. The following list provides details about each property:

1. **Agent :** This property tells the server about the client's software, usually a browser. This information is a line of text, but there is no standard format to this text. The same browser always presents the same information. For example, if you use Netscape Navigator 2.0 and your friend uses Netscape Navigator 2.0, the same information is sent by either of you when you access this server.

2. **Ip :** The ip property provides the IP address of the client. This tells the server from where on the Internet the request is sent and where to return a response. IP addresses are composed of a set of four numbers. Each is a value between 0 and 255. A sample IP number is 127.0.0.1.

   Usually the IP address you use when connecting with a host is unique, and no one else on the Net can use that address until you complete your session. However, there are special cases of caches, proxies, and firewalls that require further study if you plan to use this property.

   IP addresses can often be used to tell which domain is connecting to the server. By using what is called *reverse lookup*, the server can often determine the domain name of the client computer. You might use this to determine how many hits are coming from America Online or any other site.

3. **Method :** This property provides the HTTP method associated with the request. There are three methods currently used: get, post, and head.

4. **Protocol :** This is a standard text indicating the level of HTTP protocol used by the client. Currently this should be HTTP/1.0 in every case.

7.16.1.2. Other *Request* Properties

A request may include additional information. Your browser's request may have resulted from submitting information from an HTML form. This information becomes a request property. It does not matter if the data is submitted with either the post or get method. The name of each element of the form becomes a property name.

For example, take a form with a text element with the name of answer. Complete the text box with a value of "59" and submit it to the server. The server receives the information in the form of "answer=59." The request object now has a property called request.answer which reflects the value of "59."

7.16.1.3. Application of *Request* Properties

These built-in properties allow the server to vary the response to the request. The agent property is commonly used to supply different HTML documents to different types of browsers. For example, an HTML page with tables might only be served to a limited number of browsers while others receive a different version of the page.

Servers may limit access to certain information to specific IP addresses. Using the ip property, the server might only provide sensitive company information to clients requesting it from known company computers. Usually you want additional security, just screening IP addresses is not enough.

If a property is added for the date and time, it might be used to serve different responses. Perhaps certain parts of your site are only accessible after hours; or perhaps the graphics change depending on the time of day; or you could send a holiday greeting.

Obviously, form data sent to a server should generate a response. Perhaps a simple "thank you" and this form information is stored on the server's hard drive. Many sites have guestbooks that work this way. Other applications process the information and return a calculated response. For example, the information can be a message to append to an existing Web page. The response might acknowledge the submission but also display the updated page.

7.16.2. Client Object

Each time a new client accesses the server application, a new client object is created. However, there are no built-in properties for the client object. If you need information retained about the client, then you create a property.

Client Properties

A very common use for a client property is a client identification number. When the client first accesses the server, he can receive a form. The server can then process this information and assign the client an identification number. This might be randomly generated or looked up from a list of previous customers.

As an example, a client can send a request. Part of the request can include a user supplied customer number in the text box named idnumber. You can then create a client property called custNo with the following line:

```
client.custNo = request.idnumber
```

Client Objects Expire :

The server overflows with client objects unless they are properly deleted. Since client objects are automatically created for each and every client to access, then there must be a mechanism for deleting them.

LiveWire automatically deletes any client object with no property values. So if you don't use a client object, you don't have to worry about deleting it. In other words, you only have to clean up after yourself.

The default expiration is ten minutes of inactivity. If the client does not send another request to the server within ten minutes of the previous request, the object expires.

Obviously, in many cases ten minutes is insufficient time and, therefore, you might want to manually control this. You can simply use the expiration statement

```
client.expiration(seconds)
```

where seconds is the number of seconds before the client expires.

Another manual control is destroy. If you no longer need the client object, simply use the statement

```
client.destroy()
```

LiveWire looks at the client that sent the request and destroys its client object. This eliminates all of the client object's properties as well.

Cookies Store Information Between Sessions

Another technique for retaining client information is cookies. The browser must support the Netscape cookie protocol. If supported, the server sends the information to the client as name/value pairs. Obviously, this increases network traffic, but can offer substantial advantages to large access servers.

There are several other techniques for maintaining this information. However, they all have limited application. Client URL encoding causes a large increase in network traffic. Using IP addresses on the server only works for clients using fixed IP addresses. This might work for some intranet applications; for general use, it is worthless.

7.17. JAVASCRIPT AND CGIS

Undoubtedly with your active interest in building Websites, you have dealt with Common Gateway Interface (CGI) scripts. Prior to JavaScript, this was the primary means of creating interactive applications. Libraries of CGI scripts include counters, e-mailers, message boards, and many other functions.

LiveWire can replace CGI programming. Instead of calling external programs, the server software runs applications that are closely integrated to it. JavaScript is the language of these applications.

Applications are developed with three tools. You build these applications using LiveWire's Site Manager. The source files for the applications are developed using the same HTML editors used to build browser JavaScript pages. The applications run in response to requests from Netscape Navigator.

7.18. EXCHANGING INFORMATION BETWEEN THE BROWSER AND SERVER

In making your application come alive, the server and browser must exchange information. The client, or browser, typically sends user responses. These can first be "polished" by JavaScript routines on the browser side. The server in turn sends data back to the browser as HTML pages.

From Client to Server : User responses are submitted just as you currently handle forms. The user completes the form and clicks the submit button. The data from the radio buttons, checkboxes, textboxes, and textarea are sent to the server. The server then places this data into the request object. Each element of the form has a corresponding property.

From Server to Client : Usually a server only returns a static page in response to a browser request. In LiveWire, the response is still a page, but the contents of the page vary. User input can result in changes to default form values, new values of hidden form elements, or direct substitutions.

A server-side JavaScript can dynamically build in the HTML code for a form element that is part of the page. As an example, you can have the following statement in a source document:

```
<INPUT TYPE="text" NAME="example" VALUE='request.agent'>
```

In this case the default value of the text is the browser agent information.

You use an identical procedure for hidden form elements. The only difference is that the type is hidden instead of text. Your client-side JavaScript can then use this value as part of any function.

Another means of making your pages come alive is to change part of the JavaScript code. When you send a page to the browser it can contain JavaScript code for the browser to execute as part of the page. There is no reason that this code has to be static.

The server can modify the page being sent to change the JavaScript code embedded in the page. The page on the server side might include

```
<SERVER>
write ("<SCRIPT>var luck = " + client.winnings + "</SCRIPT>")
</SERVER>
```

Assuming the value of client.winnings is 1000, the browser sees a line of

```
<SCRIPT>var luck =  1000 </SCRIPT>
```

SUMMARY

- Netscape was the first to bring out a programming language that would allow web pages to become interactive - they called it **LiveScript**
- JavaScript is most commonly used as a client side scripting language. This means that JavaScript code is written into an HTML page. When a user requests an HTML page with JavaScript in it, the script is sent to the browser and it's up to the browser to do something with it.
- **JavaScript is *not* the same as Java.** Although the names are much alike, JavaScript is primarily a scripting language for use within HTML pages, while Java is a real programming language that does quite different things from JavaScript.
- Client and server script are generally independent. When a page is first requested, the server can run server-side script, and then pass the page to the browser, which can then run client-side script. However, if a server-side script on the page must be run again, the page must be submitted back to the server, which effectively restarts the page.
- **With Client-Side Scripting,** the scripts on the page are processed by the individual web browser that requests the page. **With Server-Side Scripting,** on the other hand, all scripts are processed on the server before the requested page is ever sent to the browser. This results in no actual code being sent to the client's machine.

SHORT QUESTION ANSWERS

Q. 1. What is JavaScript?

Ans. JavaScript is a simple scripting language invented specifically for use in web browsers to make websites more dynamic. JavaScript is an easy-to-use programming language that can be embedded in the header of your web pages. It can enhance the dynamics and interactive features of your page by allowing you to perform calculations, check forms, write interactive games, add special effects, customize graphics selections, and create security passwords and more.

Q. 2. What do you understand by JavaScript Literals?

Ans. You use literals to represent values in JavaScript. These are fixed values, not variables that you literally provide in your script. Examples of literals include: 1234, "This is a literal," and true.

Q. 3. Define the Functions in JavaScript?

Ans. Functions are one of the fundamental building blocks in JavaScript. A function is a JavaScript procedure — a set of statements that performs a specific task. A function definition has these basic parts:

- The function keyword
- A function name
- A comma-separated list of arguments to the function in parentheses
- The statements in the function in curly braces: { }

Q. 4. What is Client-Side Scripting?

Ans. Most web applications involve both client-side and server-side script. Client-side script is often used to program the user interface for an application - for example, to change a web page's text dynamically, respond to user actions such as button clicks, and perform such client-oriented tasks as input validation. Client-side script executes locally in the browser, which provides the user with a lively and responsive interface.

Q. 5. Define forms in JavaScript?

Ans. Forms are used on the web to collect information from the user. This can be anything from a Google search box through to an online application for a financial product. They take hyperlinks to a higher level, by allowing different things to happen depending on what information the form captures. A simple example is Google - entering two different keywords into the search textfield produces two different lists of returned results.

Q. 6. Explain the advantage of JavaScript on the Server?

Ans. JavaScript extends the capabilities of the server. By providing a scripting language the software can do more without calling an external program. This makes it easier for Webmasters to add features to their sites that browsers can take advantage of. And it can reduce the load on the server by keeping the processing within the server software.

REVIEW QUESTIONS

1. What is Javascript ?
2. How Javascript is different from java ?
3. Write the advantages and disadvantages of Javascript.
4. Write a program in javascript to print "Happy New Year 2010" on the screen.
5. Write a program to print 1 to 20 numbers.
6. Write the difference between client side and server side scripting.
7. Write a program to print odd numbers between 1 to 50.

❑❑❑

CHAPTER 8

Servers

8.1. INTRODUCTION TO SERVERS

INSIDE THIS CHAPTER

Client/Server is definitely a popular buzz word ? Client/Server technology is a means for separating the functions of an application into two or more distinct parts. The client presents and manipulates data on the desktop computer. The server acts like a mainframe to store and retrieve protected data. Together each machine can perform the duties it is best at.

The following are what are considered to be the benefits of client/server computing.

- Many times easier to implement client/server than change a legacy application.
- New technology and the move to rapid application development such as object oriented technology
- Long-term cost benefits for development and support.
- Easy to add new hardware to support new systems such as document imaging and video teleconferencing which would not be feasible or cost efficient in a mainframe environment.
- Can implement multiple vendor software tools for each application.

Now Let us see the definitions of server in different prospective as follows :

Definition

"A server is computer, or a software package, that provides a specific kind of service to client software running on other computers. The term can refer to a particular piece of software, such as a WWW server, or to the machine on which the software is running".

Definition

"A server is a device with a particular set of programs or protocols that provide various services, which other machines or clients request, to perform certain tasks. Together, a server and its clients form a client/server network which provides routing systems and centralized access to information, resources, stored data, etc.".

Definition

"In the client/server programming model, a server is a program that awaits and fulfills requests from client programs in the same or other computers. A given application in a computer may function as a client with requests for services from other programs and also as a server of requests from other programs".

8.1.1. Types of Servers

Servers are often dedicated, meaning that they perform no other tasks besides their server tasks. On multiprocessing operating systems however, a single computer can execute several programs at once. A server in this case could refer to the program that is managing resources rather than the entire computer. There are several types of server as follows:

1. Application Servers

Sometimes referred to as a type of middleware, application servers occupy a large chunk of computing territory between database servers and the end user, and they often connect the two.

Middleware is a software that connects two otherwise separate applications For example, there are a number of middleware products that link a database system to a Web server. This allows users to request data from the database using forms displayed on a Web browser and it enables the Web server to return dynamic Web pages based on the user's requests and profile.

2. Audio/Video Servers

Audio/Video servers bring multimedia capabilities to Websites by enabling them to broadcast streaming multimedia content. Streaming is a technique for transferring data such that it can be processed as a steady and continuous stream. Streaming technologies are becoming increasingly important with the growth of the Internet because most users do not have fast enough access to download large multimedia files quickly. With streaming, the client browser or plug-in can starts displaying the data before the entire file has been transmitted.

3. Chat Servers

Chat servers enable a large number of users to exchange information in an environment similar to Internet newsgroups that offer real-time discussion capabilities. Real time means occurring immediately.

4. Fax Servers

A fax server is an ideal solution for organizations looking to reduce incoming and outgoing telephone resources but that need to fax actual documents.

5. FTP Servers

One of the oldest of the Internet services, File Transfer Protocol makes it possible to move one or more files securely between computers while providing file security and organization as well as transfer control.

6. Groupware Servers

A GroupWare server is software designed to enable users to collaborate, regardless of location, via the Internet or a corporate Intranet and to work together in a virtual atmosphere.

7. IRC Servers

An option for those seeking real-time capabilities, Internet Relay Chat consists of various separate networks (or "nets") of servers that allow users to connect to each other via an IRC network.

8. List Servers

List servers offer a way to better manage mailing lists, whether they are interactive discussions open to the public or one-way lists that deliver announcements, newsletters, or advertising.

9. Mail Servers

Almost as ubiquitous and crucial as Web servers, mail servers move and store mail over corporate networks via LANs and WANs and across the Internet.

10. News Servers

News servers act as a distribution and delivery source for the thousands of public news groups currently accessible over the USENET news network. USENET is a worldwide bulletin board system that can be accessed through the Internet or through many online services.

11. Proxy Servers

Proxy servers sit between a client program typically a Web browser and an external server (typically another server on the Web) to filter requests, improve performance, and share connections.

12. Telnet Servers

A Telnet server enables users to log on to a host computer and perform tasks as if they're working on the remote computer itself.

13. Web Servers

At its core, a Web server serves static content to a Web browser by loading a file from a disk and serving it across the network to a user's Web browser. The browser and server talking to each other using HTTP mediate this entire exchange.

8.2. WEB SERVER [MDU May 2006, 2008]

Definition ***Web server is the program resides on server computer to server the request of the client software.***

Web servers are a combination of software and hardware that store documents and other content. The server content is accessed using Web browsers. Web servers are available for a wide range of platforms including IBM PCs, Apple Macintosh computers, UNIX platforms and a large number of proprietary hosts.

Some the example of the web servers are as follows:

1. Tomcat
2. Weblogic
3. Apache
4. IIS.

Let's say that we are sitting at our computer, surfing the Web, and we get a call from a friend who says, "I just read a great article! Type in this URL and check it out. It's at http://www.kiet.edu/adesh.htm." So we type that URL into our browser and press return. And magically, no matter where in the world that URL lives, the page pops up on our screen.

At the most basic level possible, the following diagram shows the steps that brought that page to our screen:

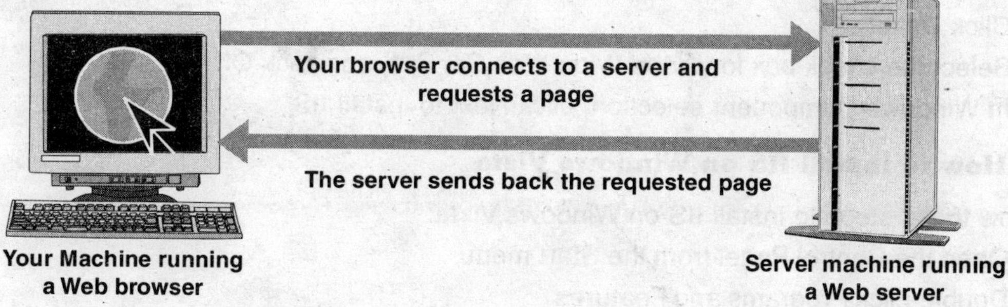

FIGURE 8.1. Functioning of Web Server.

Our browser formed a connection to a Web server, requested a page and received it.

8.2.1. Working of Web Servers

Let us know the working of web servers after knowing the definition of server and web servers. Here are the basic steps that occurred behind the scenes when we try to get particular web page on our computer:

Step 1 : The browser broke the URL into three parts:

1. The protocol ("http")
2. The server name ("www.kiet.edu")
3. The file name ("adesh.htm")

Step 2 : The browser communicated with a name server to translate the server name "www.kiet.edu" into an **IP Address**, which it uses to connect to the server machine.

Step 3 : The browser then formed a connection to the server at that IP address on port 80.

Step 4 : Following the HTTP protocol, the browser sent a GET request to the server, asking for the file "http://www.kiet.edu/adesh.htm."

Step 5 : The server then sent the HTML text for the Web page to the browser.

Step 6 : The browser read the HTML tags and formatted the page onto your screen.

8.3. INTERNET INFORMATION SERVER (IIS) [MDU May 2006]

Definition

IIS is a set of Internet-based services for servers created by Microsoft for use with Microsoft Windows.IIS comes with Windows 2000, XP, and Vista. It is also available for Windows NT. IIS is easy to install and ideal for developing and testing web applications.

IIS includes Active Server Pages (ASP), a server-side scripting standard that can be used to create dynamic and interactive web applications. It is the world's second most popular web server in terms of overall websites behind the industry leader Apache HTTP Server.

8.3.1. How to Install IIS on Windows XP and Windows 2000

Follow these steps to install IIS on Windows XP, and Windows 2000:

1. On the Start menu, click Settings and select Control Panel
2. Double-click Add or Remove Programs
3. Click Add/Remove Windows Components
4. Click Internet Information Services (IIS)
5. Click Details
6. Select the check box for World Wide Web Service, and click OK
7. In Windows Component selection, click Next to install IIS

8.3.2. How to Install IIS on Windows Vista

Follow these steps to install IIS on Windows Vista:

1. Open the Control Panel from the Start menu
2. Double-click Programs and Features
3. Click "Turn Windows features on or off" (a link to the left)
4. Select the check box for Internet Information Services (IIS), and click OK

8.3.3. Test Your Web Page [MDU May 2006]

After you have installed IIS or PWS follow these steps:

1. Look for a new folder called **Inetpub** on your hard drive
2. Open the Inetpub folder, and find a folder named **wwwroot**
3. Create a new folder, like "MyWeb", under wwwroot
4. Write some ASP code and save the file as "test1.asp" in the new folder
5. Make sure your Web server is running (see below)

Open your browser and type "http://localhost/MyWeb/test1.asp", to view your first web page

8.3.4. Creating New Web Sites on IIS

Let's suppose your web application consists of a E-Commerce website, called example.com. This ASP.NET web application has a main user interface for registered users and guests as well as an Administration Panel. This panel will be served from a different secure location installed on the target server, hence the need to configure a separate virtual directory for this application.

1. Add your application files
2. Add your Web Site
3. Configure Access Flags for your Web Site
4. Set up default documents for your application
5. Enable FTP access
6. Configure Virtual Directory
7. Configure global IIS settings (optional)
8. Run the project

1. Add your Application Files

Go to the Files and Folders view and add your application files. The application's files for the Administration Panel, which will be handled by the virtual directory adminpanel of the website, should be placed in a separate directory.

It is redundant to add as a virtual directory a child folder of your application's website folder, because this would already be served by your website.

2. Add your Web Site

Use the [New Web Site] toolbar button to add your website called example.com. Configure it as follows:

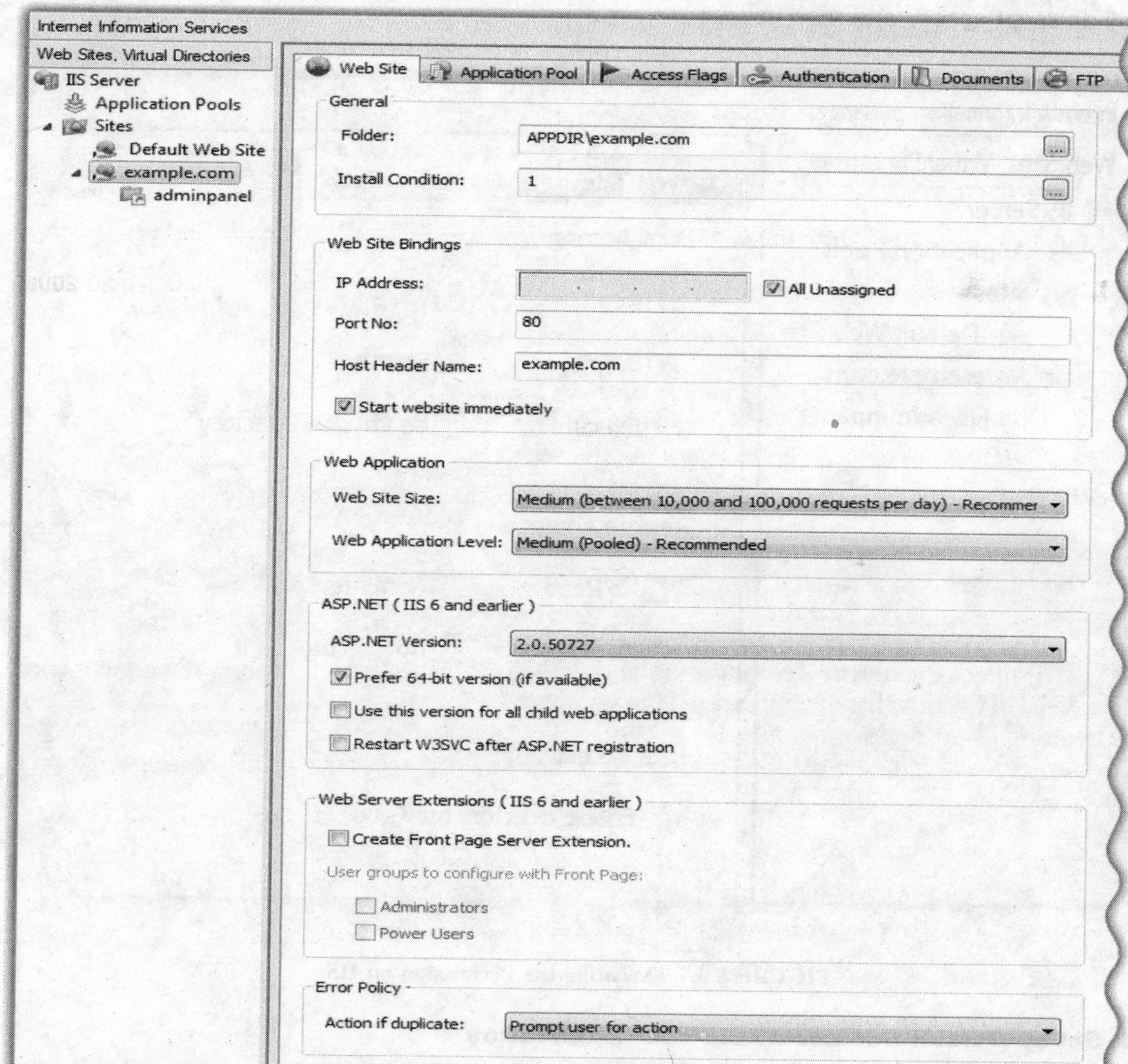

FIGURE 8.2. **Add your Web Site on IIS.**

This triplet setting (IP Address, Port No, Host Header Name) defines the Web Site and therefore must be unique.

If you build and maintain Websites for customers and coordinate these efforts with multiple authors, then the features included in FrontPage Server Extensions can help you accomplish this task. FrontPage Server Extensions provide Web-based and command-line administration for extending virtual servers. Extending virtual servers with FrontPage Server Extensions enables the site owner to author the site in FrontPage and delegate site ownership and administration credentials.

3. Configure Access Flags for your Web Site

In our scenario, the ASP.NET web application requires **execute permission** for the scripts as well as script access and read access.

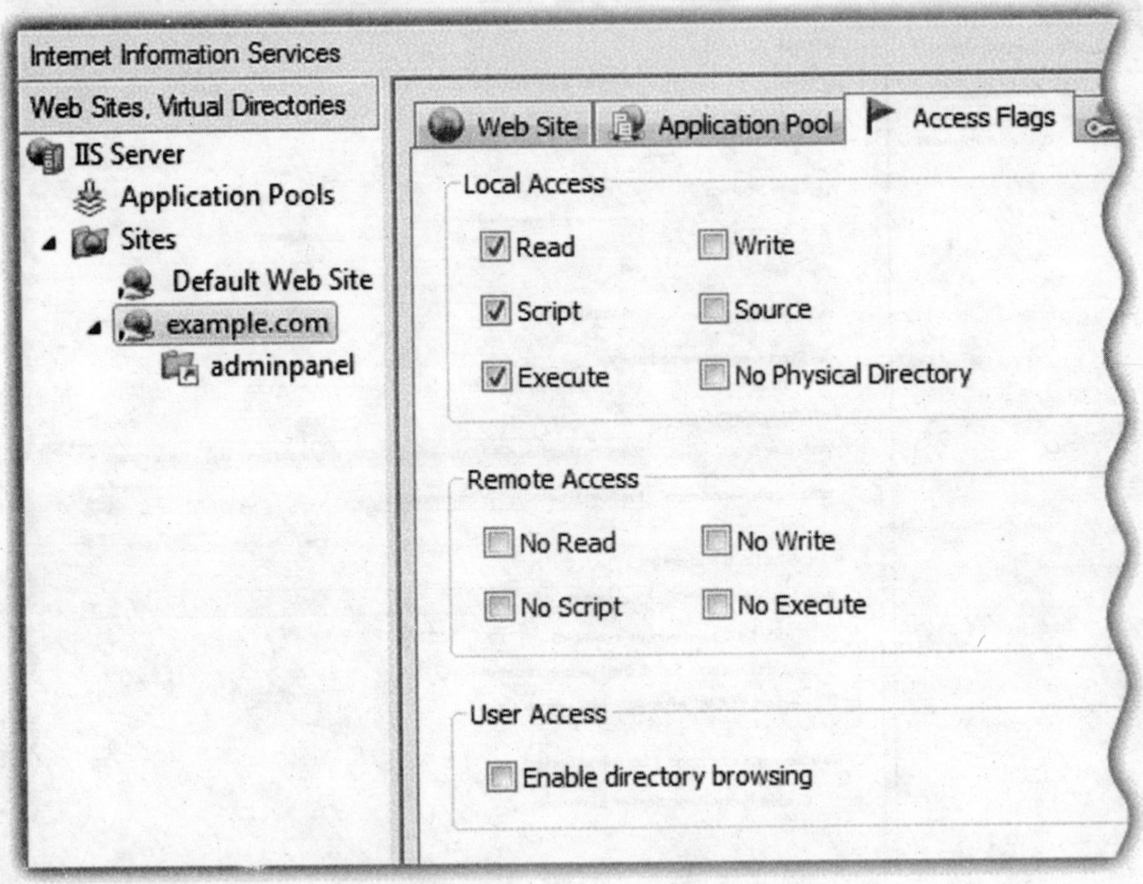

FIGURE 8.3. Executing the Permission on IIS.

4. Set up Default Documents for Your Application

Our Web Site will automatically serve a default document when no document name is specified in the request (url). In our case index.aspx will have priority whenever found in a directory.

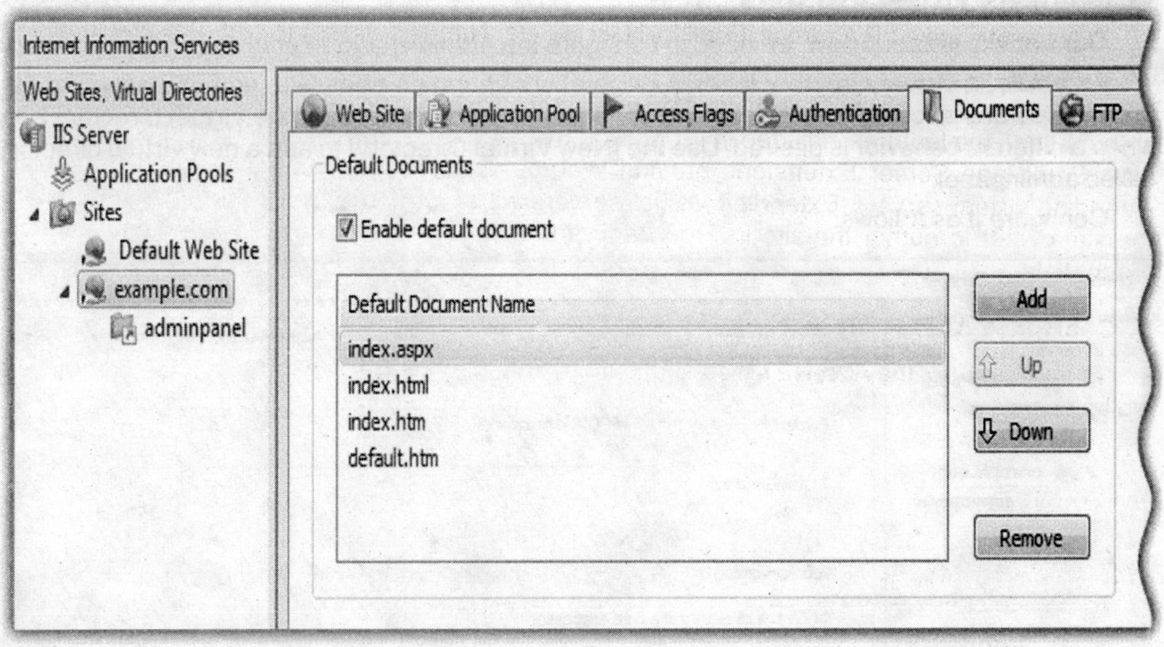

FIGURE 8.4. **Setting the Default Page on IIS.**

5. Enable FTP Access

Our application requires FTP access for constant updating of script content, so we will configure FTP access for the web application. We do not set the "Set NTFS security flags" option to allow only registered users on the target machine, having the proper access rights, to connect through FTP.

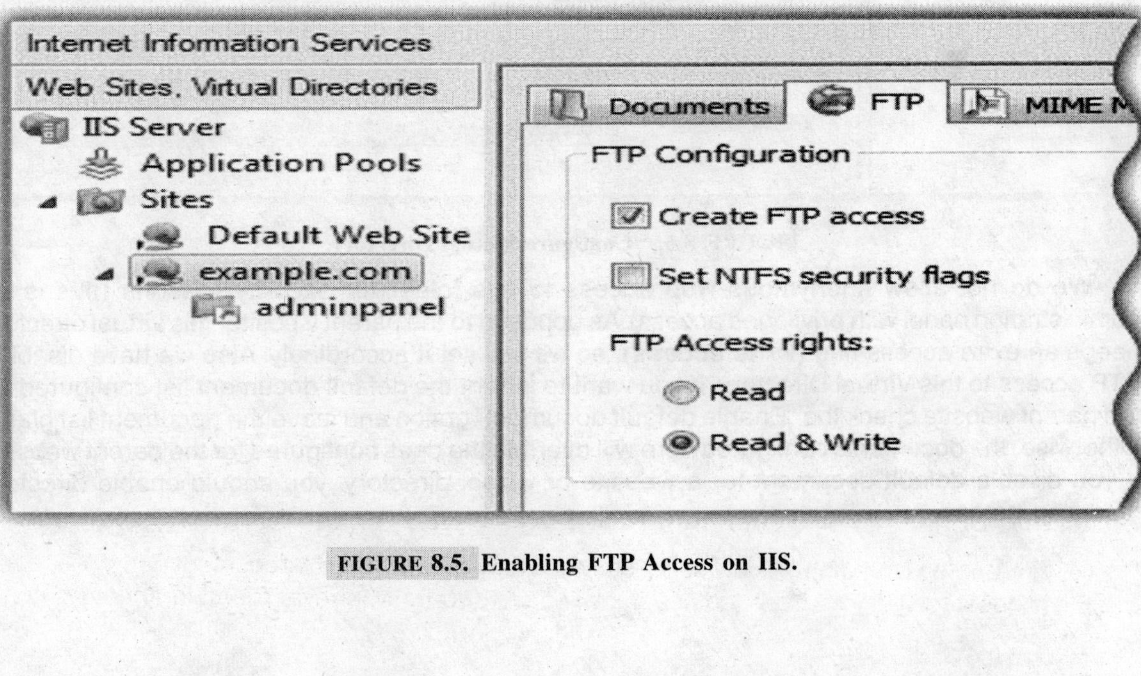

FIGURE 8.5. **Enabling FTP Access on IIS.**

6. Configure Virtual Directory

Our website is set up now, we need to configure the Administration Panel as a virtual directory.

As a rule, a virtual directory inherits general configuration traits like: default document list, mime types and access flags; so, we will configure such properties for the virtual directory only when a different behavior is desired. Use the [New Virtual Directory] to add a new virtual directory called adminpanel.

Configure it as follows:

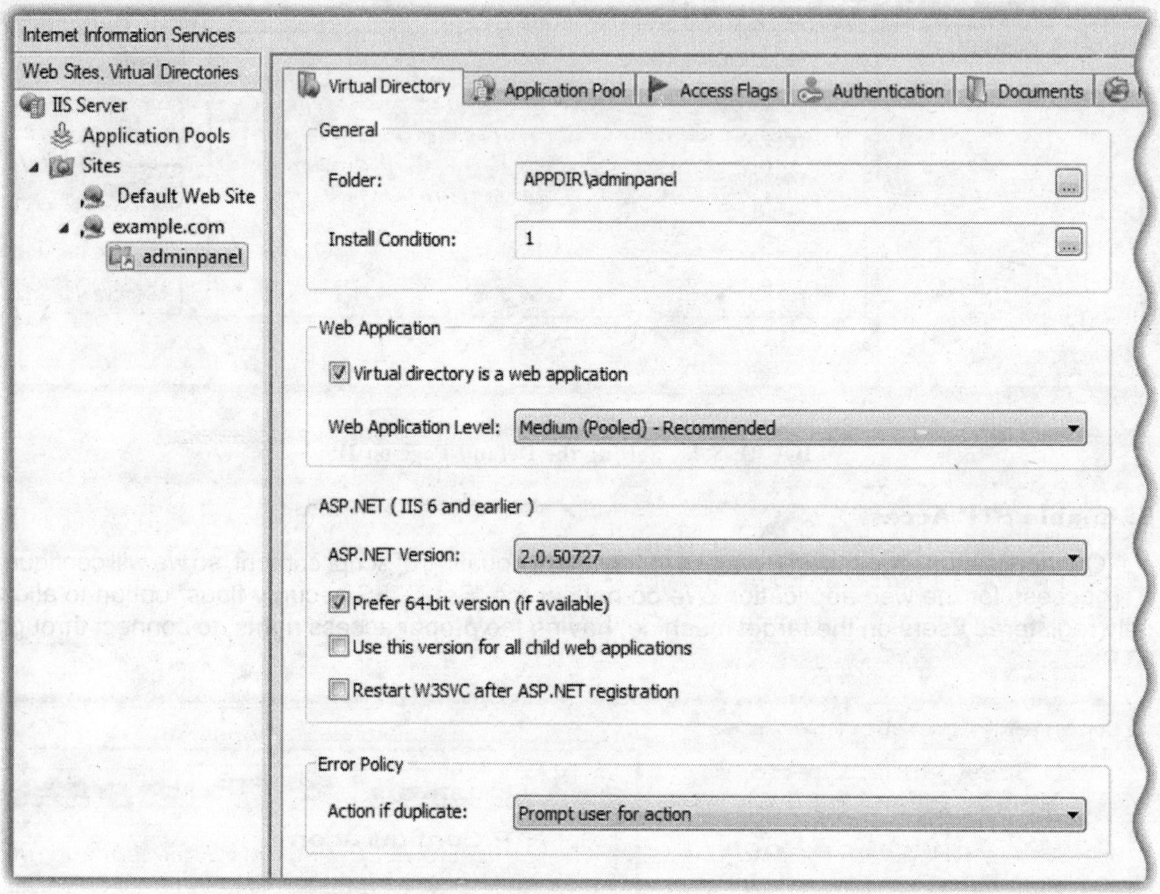

FIGURE 8.6. Configure Virtual Directory.

We do not allow anonymous web access to this folder for security reasons (this is an administration panel with privileged access).As opposed to the parent website, this virtual directory needs an extra access flag (**write access**), so we will set it accordingly. Also we have disabled FTP access to this Virtual Directory. If you want to inherit the default document list configured for the parent website check the "Enable default document" option and leave the document list blank. Otherwise, the documents configured here will override the ones configured for the parent website. If you disable default document for a website or virtual directory, you should enable directory browsing instead.

7. Configure Global IIS Settings (Optional)

If needed, you can also configure global IIS settings that will affect all web sites and virtual directories, such as: MIME types, ISAPI filters.

8. Run the Project

Our website configuration is complete, build and run the package.

8.4. PERSONAL WEB SERVER (PWS) [MDU Dec. 2007]

Definition

PWS, an abbreviation for Personal Web Server, is Microsoft's version of a Web server program for individual PC users who want to share Web pages and other files from their hard drive. PWS is a scaled-down version of Microsoft's more robust Web server, Internet Information Server IIS.

PWS can be used with a full-time Internet connection to serve Web pages for a Website with limited traffic. It can also be used for testing a Website offline or from a "staging" site before putting it on a main Website that is exposed to larger traffic.

Here's what the Personal Web Server does for you:

- Allows webs to be published as a whole, with a single command.
- Publish from local to remote, remote to local, or remote to remote websites.
- Provides the functionality of CGI, IDC, ASP, ISAPI programs, and the FrontPage WebBots on your local computer, so that you can test your webs locally in your browser before making them public.
- Provides "smart" web management.
- Set permissions and properties of webs and folders.
- Sets up virtual directories on the server.
- You can actually browse your webs from another computer on the Internet, when you are logged on. This can be handy for demonstrating them to clients, friends, etc.

8.4.1. How to Install PWS on Windows 95, 98, and Windows NT

For Windows 98: Open the **Add-ons** folder on your Windows CD, find the **PWS** folder and run **setup.exe** to install PWS.

For Windows 95 or Windows NT: Download "Windows NT 4.0 Option Pack" from Microsoft, and install PWS.

Test Your Web

After you have installed IIS or PWS follow these steps:

1. Look for a new folder called **Inetpub** on your hard drive
2. Open the Inetpub folder, and find a folder named **wwwroot**
3. Create a new folder, like "MyWeb", under wwwroot
4. Write some ASP code and save the file as "test1.asp" in the new folder
5. Make sure your Web server is running (see below)

Open your browser and type "http://localhost/MyWeb/test1.asp", to view your first web page

8.5. APACHE

Definition

The Apache HTTP Server Project is a collaborative software development effort aimed at creating a robust, commercial-grade, and freely-available source code implementation of an HTTP (Web) server.

The project is jointly managed by a group of volunteers located around the world, using the Internet and the Web to communicate, plan, and develop the server and its related documentation. This project is part of the Apache Software Foundation. In addition, hundreds of users have contributed ideas, code, and documentation to the project. This file is intended to briefly describe the history of the Apache HTTP Server and recognize the many contributors.

TIPS

Apache is generally recognized as the world's most popular Web server (HTTP server). Apache has attracted so much interest because it is full-featured, reliable, and free. Originally developed for UNIX operating systems, Apache has been updated to run on Windows, OS/2, and other platforms.

8.5.1. Features of Apache Web Server [MDU Dec. 2006, May 2007]

The features of this server are as follows :

1. Is a powerful, flexible, HTTP/1.1 compliant web server
2. Implements the latest protocols, including HTTP/1.1 (RFC2616)
3. Is highly configurable and extensible with third-party modules
4. Can be customized by writing 'modules' using the Apache module API
5. Provides full source code and comes with an unrestrictive license
6. Runs on Windows 2003/XP/2000/NT/9x, Netware 5.x and above, OS/2, and most versions of Unix, as well as several other operating systems
7. Is actively being developed
8. Encourages user feedback through new ideas, bug reports and patches

8.5.2. Installation of Apache Web Server

We should follow the following steps to install the Apache Web Server:

1. Download Apache

I recommend downloading the latest stable release. At the time of this writing, that was Apache 2.0. The best place to get Apache is from the Apache HTTP Server download site. Download the binary for Windows32.

2. Extract the Files

Once you've downloaded the binary, run it to start the self-installation. Here is a step-by-step screen shot demo of what to expect.

3. Double-Click the Apache Executable and Get the Welcome Screen

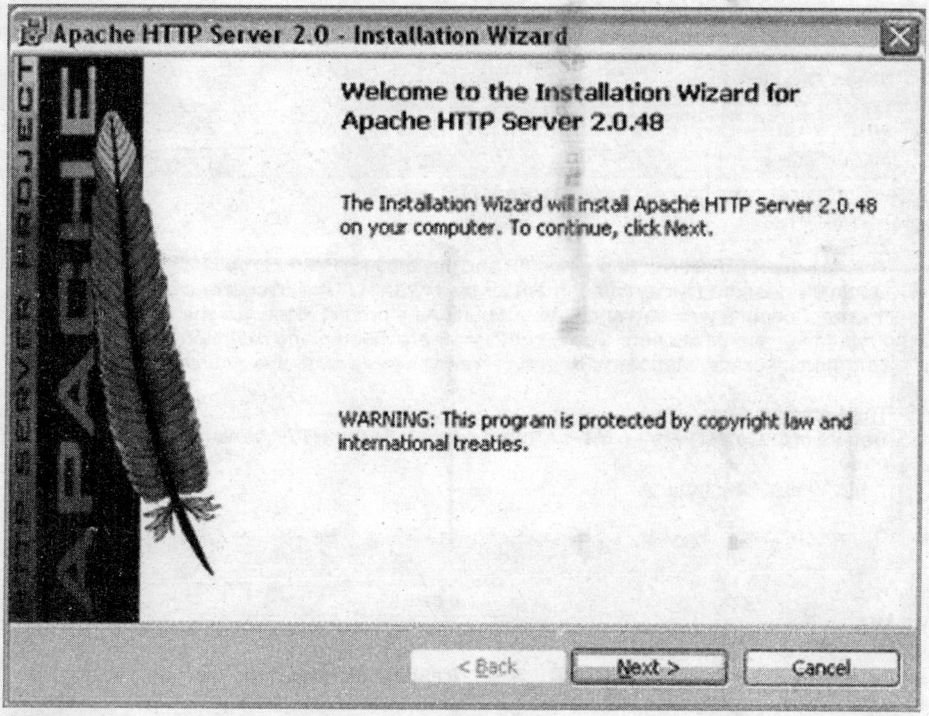

FIGURE 8.7. Welcome Screen.

4. Read the License

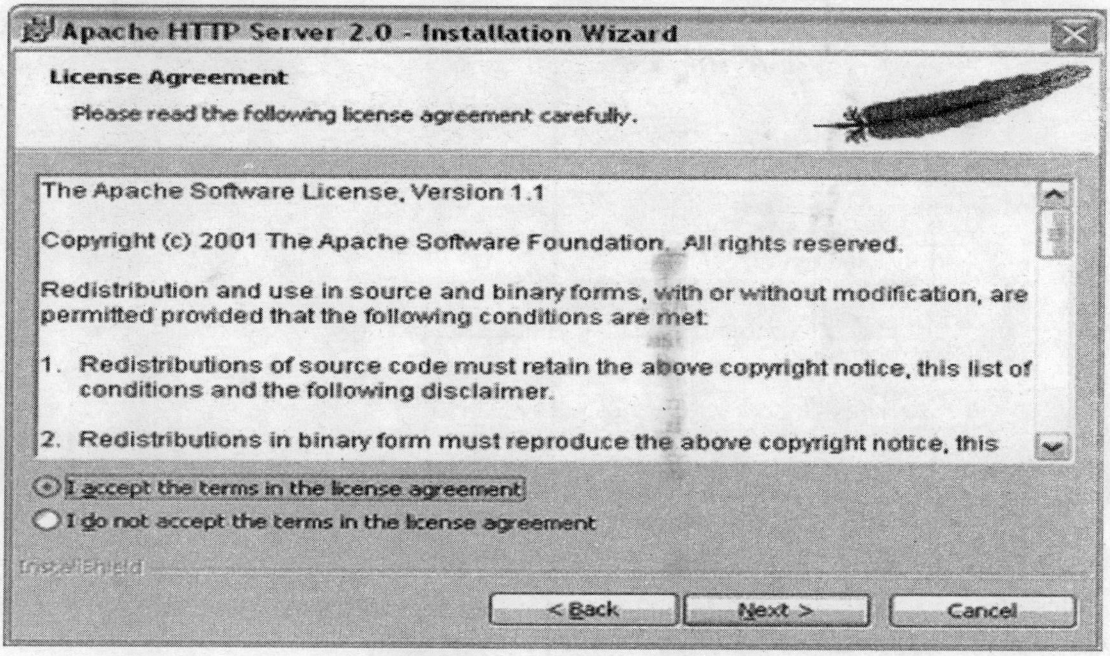

FIGURE 8.8. License Screen.

5. Read the README File

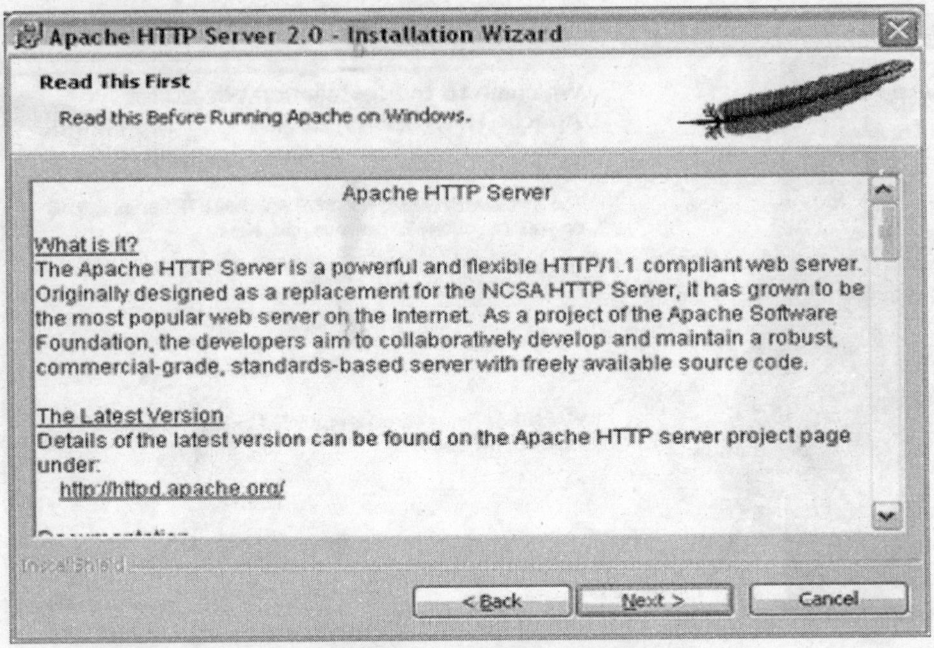

FIGURE 8.9. Readme Screen.

6. Provide Your Server Information

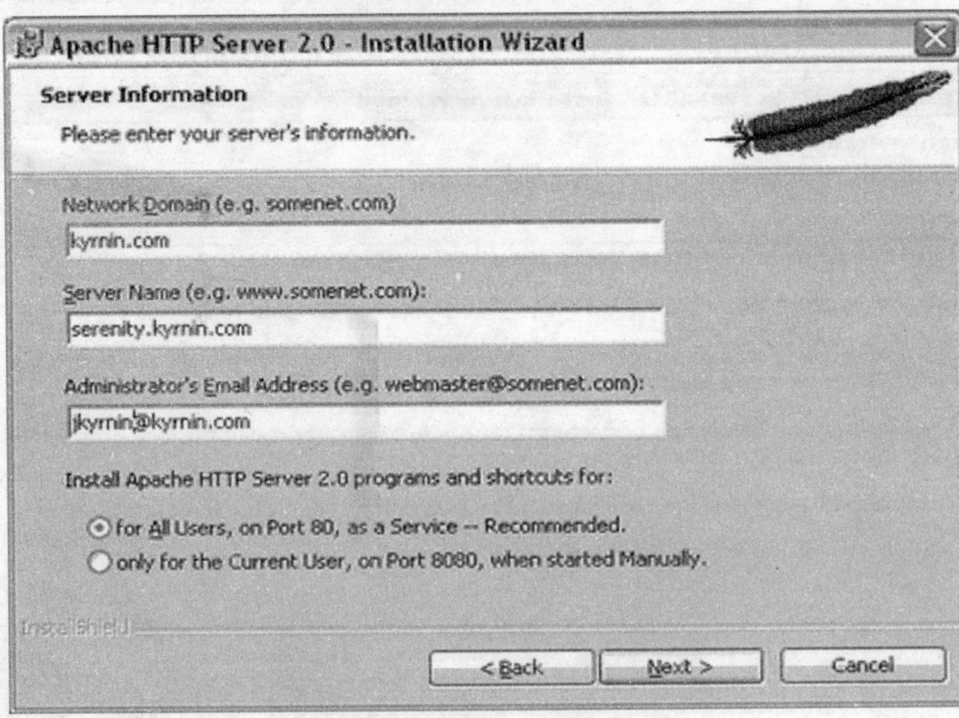

FIGURE 8.10. Information Screen.

7. Setup Type

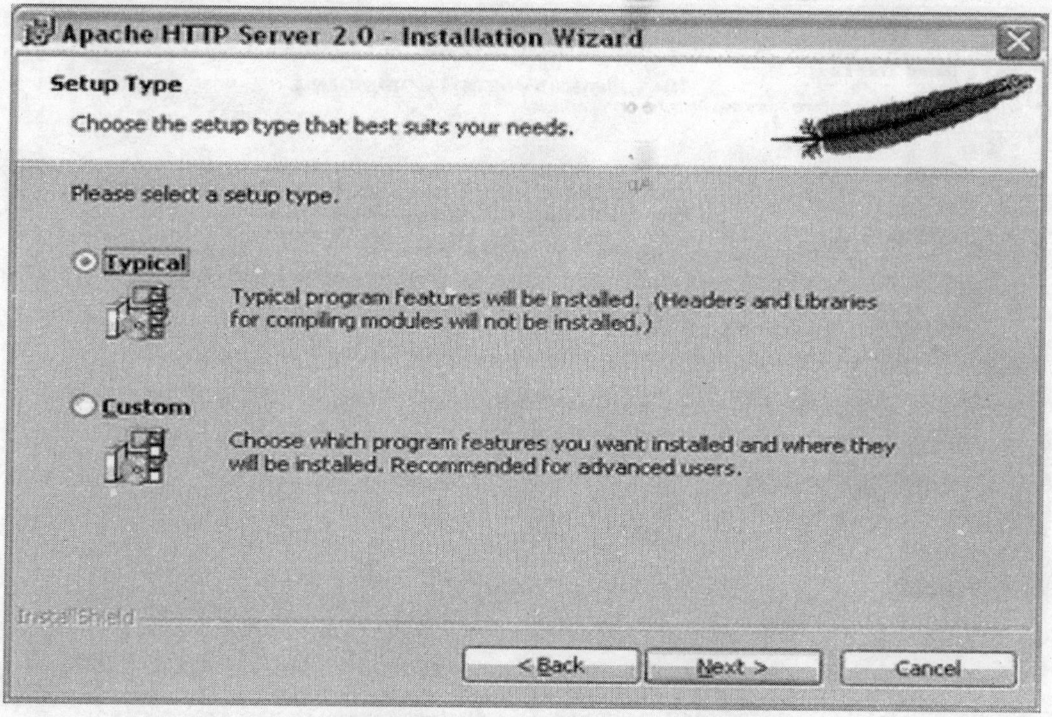

FIGURE 8.11. Setup Type Screen.

8. Review Custom Options

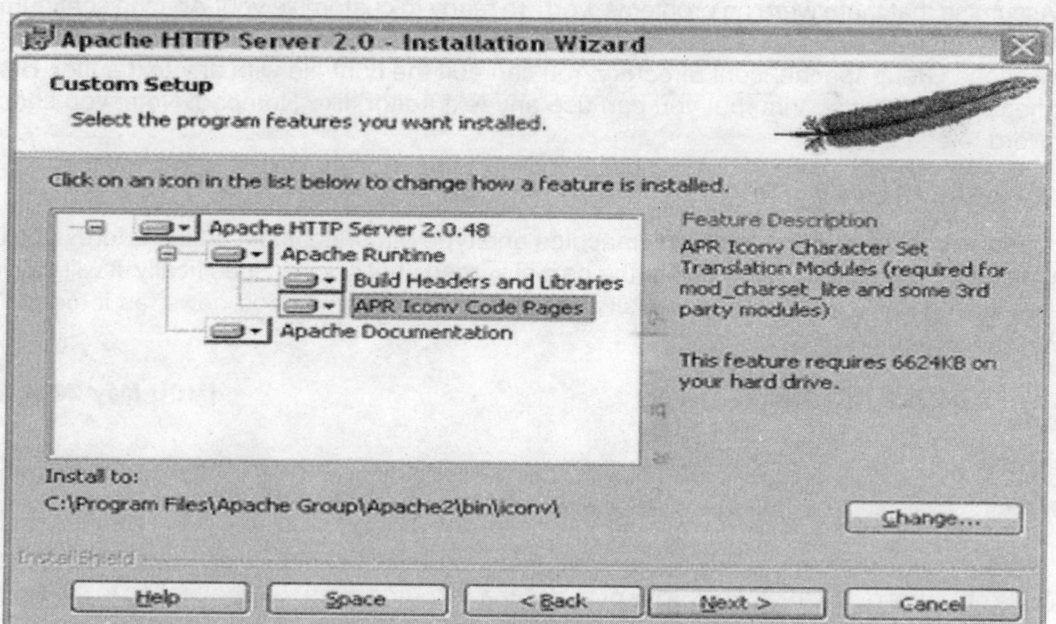

FIGURE 8.12. Custom Screen.

9. You're Done

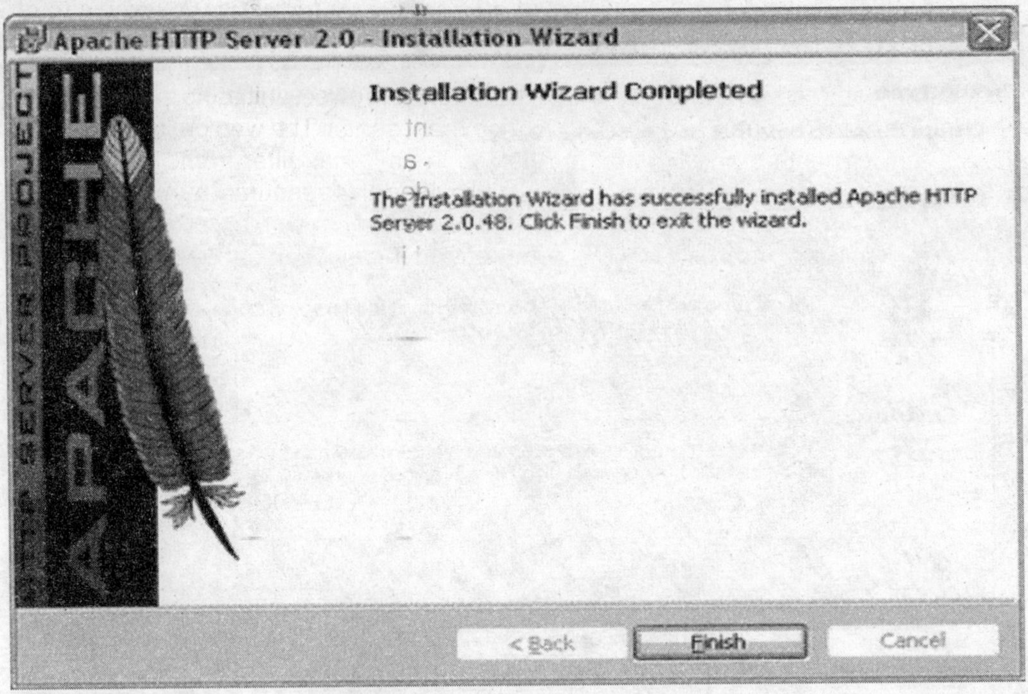

FIGURE 8.13. **Completion Screen.**

10. Customize

Assuming that there were no problems, you are ready to customize your Apache configuration. This really just amounts to editing the httpd.conf file. This file is located in the C:Program FilesApache GroupApache2conf directory. You can edit the conf file with any text editor. Being a unix-head, I usually use Vim, but you can use any text editor, like Notepad. Note: you shouldn't use Word to edit this file.

11. Test Your Server

Open a Web browser on the same machine and type http://localhost/ in the address box. You should see a page similar to the one in the partial screen shot above. Specifically, it will say in big letters "Seeing this instead of the website you expected?" This is good news, as it means your server is installed correctly.

8.6. PROXY SERVER [MDU May 2005, 2007]

Definition *"A proxy server is a server that acts between user's networks and the Internet for seeking resources from other servers".*

A client connects to the proxy server, and then requests a connection, file, or other resource available on a different server. The proxy provides the resource either by connecting to the specified server or by serving it from a cache. In some cases, the proxy may alter the client's request or the server's response for various purposes.

Definition *"A proxy server is a computer that offers a computer network service to allow clients to make indirect network connections to other network services".*

A proxy users connects to the proxy server through a browser initiating access request to some web page, or other resource, available from a different server. The web proxy server analyzes and evaluates the request as per its filtering rules such as it may filter traffic by IP address or protocol. If the request gets validated, the proxy provides the resource by connecting to the relevant server and requesting the service on behalf of the user. A web proxy server may modify or adjust the users request or the server's response, and it may even serve the request without contacting the specified server.

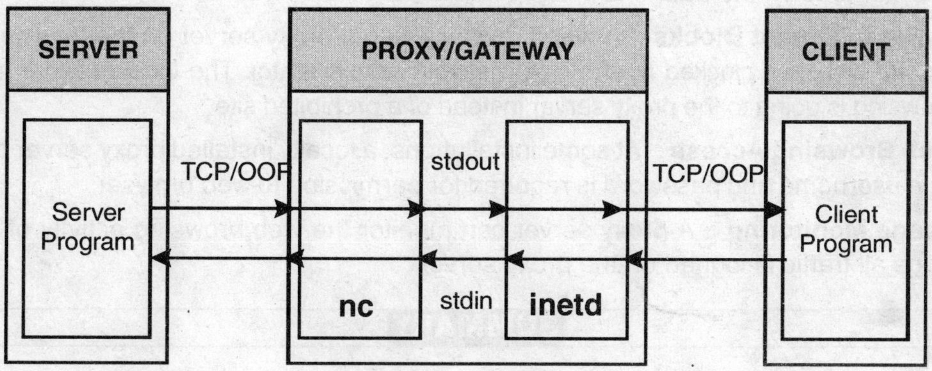

FIGURE 8.14. Proxy Server.

The purpose of a proxy server is to keep the machines unidentified for security reason and to expedite access to a resource by caching web pages from the web server. A proxy server can be located in the user's local computer or at diverse points between the user and the destination servers or the Internet. A reverse proxy is a proxy used as a front-end to accelerate and cache in-demand resources like the web pages.

8.6.1. Types of Proxy Servers

There are four different types of proxy servers :

1. **Transparent Proxy :** This type of proxy server identifies itself as a proxy server and also makes the original IP address available through the http headers. These are generally used for their ability to cache websites and do not effectively provide any anonymity to those who use them. They are transparent in the terms that your IP address is exposed, not transparent in the terms that you do not know that you are using it (your system is not specifically configured to use it.)
2. **Anonymous Proxy :** This type of proxy server identifies itself as a proxy server, but does not make the original IP address available. This type of proxy server is detectable, but provides reasonable anonymity for most users.
3. **Distorting Proxy :** This type of proxy server identifies itself as a proxy server, but make an incorrect original IP address available through the http headers.
4. **High Anonymity Proxy :** This type of proxy server does not identify itself as a proxy server and does not make available the original IP address.

8.6.2. Purpose of a Proxy Server

A proxy server acts as a go-between for web browsers and the websites they access. When a web browser is configured to use a proxy server, all traffic from the web browser goes to the proxy server, which forwards traffic to the final destination. Both users and administrators have many reasons to use a proxy server.

1. **Privacy :** An anonymous proxy server hides the originating IP of the web browser from the website. The website will see the IP of the proxy server instead of the user.
2. **Caching :** Proxy servers can cache data and images locally, thus allowing the web browser to fetch the data and images faster from slow websites.
3. **Bypass Content Blocks :** A web browser using a proxy server on the Internet can get around websites blocked by the local network administrator. The local network thinks the browsing is going to the proxy server instead of a prohibited site.
4. **Web Browsing Access :** At some installations, a locally installed proxy server protected by a username and password is required for permission to web browser.
5. **Usage Monitoring :** A proxy server can monitor the web browsing activity of its users since all traffic is logged on the proxy server.

SUMMARY

- This is the age of Internet and the complete internet is based on client-server technology. The client and server computer are two most important members of client-server technology.
- The students will gain the basic knowledge about server and their workings by going through this chapter. Students will get knowledge about popular servers like IIS, PWS, and Apache.
- Client/Server technology is a means for separating the functions of an application into two or more distinct parts. The client presents and manipulates data on the desktop computer. The server acts like a mainframe to store and retrieve protected data.
- Web servers are a combination of software and hardware that store documents and other content. The server content is accessed using Web browsers.
- Web servers are available for a wide range of platforms including IBM PCs, Apple Macintosh computers, UNIX platforms and a large number of proprietary hosts.
- A client connects to the proxy server, and then requests a connection, file, or other resource available on a different server. The proxy provides the resource either by connecting to the specified server or by serving it from a cache. In some cases, the proxy may alter the client's request or the server's response for various purposes.

SHORT QUESTION ANSWERS

Q. 1. Define the server computer?

Ans. The A server is computer, or a software package, that provides a specific kind of service to client software running on other computers. The term can refer to a particular piece of software, such as a WWW server, or to the machine on which the software is running".

Q. 2. What is Application Servers?

Ans. Sometimes referred to as a type of middleware, application servers occupy a large chunk of computing territory between database servers and the end user, and they often connect the two. Middleware is a software that connects two otherwise separate applications

Q. 3. Define Chat Servers?

Ans. Chat servers enable a large number of users to exchange information in an environment similar to Internet newsgroups that offer real-time discussion capabilities. Real time means occurring immediately.

Q. 4. What do you understand by FTP Server?

Ans. One of the oldest of the Internet services, File Transfer Protocol makes it possible to move one or more files securely between computers while providing file security and organization as well as transfer control.

Q. 5. What is Mail Servers?

Ans. Almost as ubiquitous and crucial as Web servers, mail servers move and store mail over corporate networks via LANs and WANs and across the Internet.

Q. 6. What is Proxy Servers?

Ans. Proxy servers sit between a client program typically a Web browser and an external server (typically another server on the Web) to filter requests, improve performance, and share connections.

Q. 7. Define Web Servers?

Ans. At its core, a Web server serves static content to a Web browser by loading a file from a disk and serving it across the network to a user's Web browser. The browser and server talking to each other using HTTP mediate this entire exchange.

Q. 8. What is Apache Server?

Ans. **Apache** is generally recognized as the world's most popular Web server (HTTP server). Apache has attracted so much interest because it is full-featured, reliable, and free. Originally developed for UNIX™ operating systems, Apache has been updated to run on Windows, OS/2, and other platforms.

Q. 9. What is Transparent Proxy?

Ans. This type of proxy server identifies itself as a proxy server and also makes the original IP address available through the http headers. These are generally used for their ability to cache websites and do not effectively provide any anonymity to those who use them. However, the use of a transparent proxy will get you around simple IP bans. They are transparent in the terms that your IP address is exposed, not transparent in the terms that you do not know that you are using it (your system is not specifically configured to use it.)

Q. 10. What is Anonymous Proxy?

Ans. This type of proxy server identifies itself as a proxy server, but does not make the original IP address available. This type of proxy server is detectable, but provides reasonable anonymity for most users.

Q. 11. What is Distorting Proxy?

Ans. This type of proxy server identifies itself as a proxy server, but make an incorrect original IP address available through the http headers.

REVIEW QUESTIONS

1. What is client server technology and how it is advantageous?
2. Define the server in different prospects.
3. Explain the different type of servers.
4. Explain the web server with examples.
5. Explain the working of web server.
6. What is IIS and how can we install it on window XP?
7. What is PWS and how can be install it?
8. What is Apache server?

❑❑❑

PRIVACY AND SECURITY TOPICS

SECTION- D

- **Web Security and Firewall**
- **Privacy and Encryption Schemes**

SYLLABUS

Introduction, Software Complexity, Encryption schemes, Secure Web document, Digital Signature, Firewalls.

9

CHAPTER

Web Security and Firewall

INSIDE THIS CHAPTER

9.1. WORLD WIDE WEB SECURITY

We know that almost all e-commerce and m-commerce activities can be performed through WWW services. Most of the transactions are done for important information and finance related. So security is the major thrust are for transaction take place on web. In order to counter the various types of security threats/attacks to transmitted electronic data, security services to ensure reliable and true worthy transmission of digital business data and messages are needed. The primary security services are divided into five categories :

- Confidentiality
- Integrity
- Authentication
- Non-repudiation
- Authorization
- Availability

Different organizations and individual may face different types of threats to the security of their information on the Internet. Some of the common risks/threats are :

1. Data may be stolen, corrupted misused, altered.
2. Attacks on hardware may occur that may render systems unable to operate properly.
3. Programs may be altered to cause systems to perform incorrectly or even to crash.

9.1.1. Types of Web Security Risks

There are basically three overlapping types of risk :

1. **Bugs or misconfiguration problems in the Web server** that allow unauthorized remote users to :

(a) Steal confidential documents not intended for their eyes.
(b) Execute commands on the server host machine, allowing them to modify the system.
(c) Gain information about the Web server's host machine that will allow them to break into the system.
(d) Launch denial-of-service attacks, rendering the machine temporarily unusable.

2. **Browser-side risks, including :**
   - Active content that crashes the browser, damages the user's system, breaches the user's privacy, or merely creates an annoyance.
   - The misuse of personal information knowingly or unkowingly provided by the end-user.
3. **Interception of network data sent from browser to server or vice versa via network hackers.** Hackers can operate from any point on the pathway between browser and server including :
   (a) The network on the browser's side of the connection.
   (b) The network on the server's side of the connection (including intranets).
   (c) The end-user's Internet Service Provider (ISP).
   (d) The server's ISP.
   (e) Either ISPs' regional access provider.

TIPS

It's important to realize that "secure" browsers and servers are only designed to protect confidential information against network hackers. Without system security on both browser and server sides, confidential documents are vulnerable to interception.

9.1.2. Web Security Precautions

If we are a Webmaster, system administrator, or are otherwise involved with the administration of a network, the single most important step we can take to increase our site's security is to create a written security policy. This security policy should succinctly lay out our organization's policies with regard to :

(a) who is allowed to use the system
(b) when they are allowed to use it
(c) what they are allowed to do (different groups may be granted different levels of access)
(d) procedures for granting access to the system
(e) procedures for revoking access (e.g., when an employee leaves)
(f) what constitutes acceptable use of the system
(g) remote and local login methods
(h) system monitoring procedures
(i) protocols for responding to suspected security breaches

This policy need not be anything fancy. It need only a succinct summary of how the information system work, reflecting your organization's technological and political realities. There are several benefits to having a written security policy:

1. We ourself will understand what is not permitted on the system. If we don't have a clear picture of what is permitted, we can never be sure when a violation has occurred.
2. Others in our organization will understand what the security policy is. The written policy raises the level of security consciousness, and provides a focal point for discussion.
3. The security policy serves as a requirements document against which technical solutions can be judged. This helps guard against the "buy first, ask questions later" syndrome.

9.2. CLIENT SERVER TECHNOLOGY

Client/Server technology is a means for separating the functions of an application into two or more distinct parts. The client presents and manipulates data on the desktop computer. The server acts like a mainframe to store and retrieve protected data. Together each machine can perform the duties it is best at.

Among the characteristics of a client/server architecture are the following :

(a) The client and server can be distinguished from one another by the differences in tasks they perform.

(b) The client and server usually operate on different computer platforms.

(c) Either the client or server may be upgraded without affecting the other. Clients may connect to one or more servers; servers may connect to multiple clients concurrently.

(d) Clients always initiate the dialogue by requesting a service.

9.2.1. Overview of the Client Server Technology

Here's a little map of Client Server Technologies. The client computer (the hardware sitting perhaps on our desktop) communicates across the Internet (by modem or Ethernet) with a server computer (usually at a remote site).

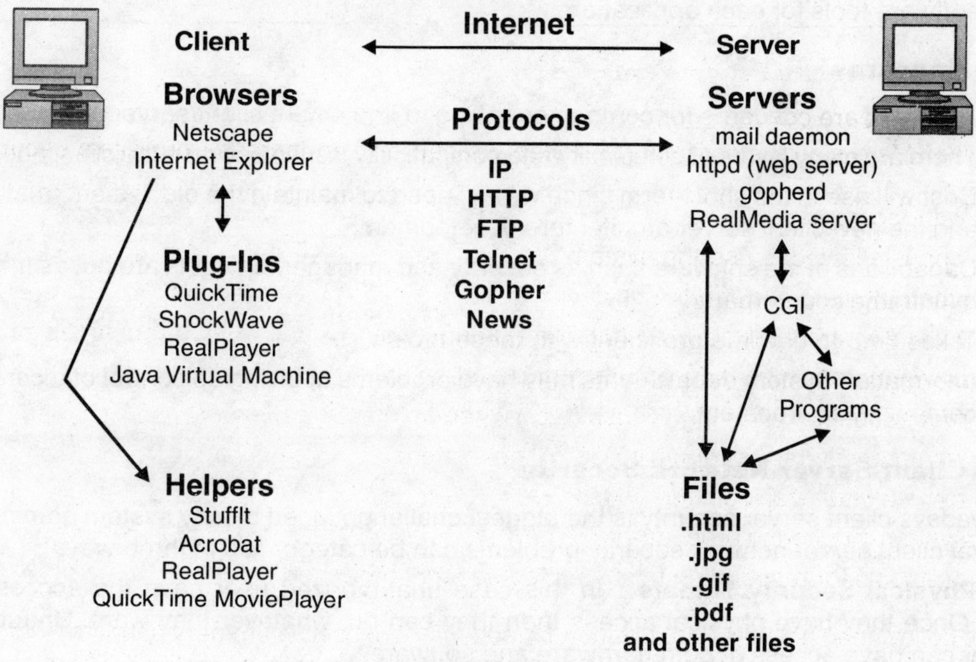

FIGURE 9.1.

The client software on the client computer is a web browser or a telnet program. It contacts software called servers on the server computer. These servers are usually daemons (a name for software that is always running). The client software and the server software communicate by means of a protocol (an established pattern of behaviours and responses).

So for example, a **web browser may contact a web server using the HTTP protocol.** The browser sends a request for a particular web page. The web server logs the request and then finds the correct file and sends it back across the Internet to the browser. The browser receives the file and interprets it for the user by supplying images, format and so on.

Helper Applications run on the client's side to help show files that require special handling, such as movies, or compressed files, or Acrobat (.pdf) files. **Plug-Ins are software that actually run inside the web brower to display special files.**

On the server side, Common Gateway Interface (CGI) programs can access various files and other programs on the server computer and send dynamically created web pages back across the Internet. For example, this is the way search engines work: each search creates a new page that doesn't already exist as a file.

9.2.2. Benefits

The following are what are considered to be the benefits of client/server computing.

(a) Many times easier to implement client/server than change a legacy application.

(b) New technology and the move to rapid application development such as object oriented technology

(c) Long-term cost benefits for development and support. Easy to add new hardware to support new systems such as document imaging and video teleconferencing which would not be feasible or cost efficient in a mainframe environment. Can implement multiple vendor software tools for each application.

9.2.3. Concerns

The following are concerns for companies willing to implement client/server computing.

(i) There are many layers of complexity and compatibility issues between the client and server.

(ii) Cost will rise in the short-term since we will need to maintain the old system (mainframe) and the new client server architecture development.

(iii) Capabilities of the software such as security and management tools are not as mature as mainframe counterparts.

(iv) Takes time to become proficient with these tools.

(v) Information System departments may have problems at giving up control of a centralized computing environment.

9.2.4. Client Server Network Security

Nowadays client server security is the biggest challenge faced by any system administrator. In general client server network security problems can be categorised in three ways :

(a) **Physical Security Threats :** In this case unauthorized user have the access of our system. Once they have physical access then they can do, whatever they want. Unauthorized accesser can have access of both hardware and software.

(b) **Software Security Threats :** In this case a poorly written program can do the damage of highest level. Through these codes intruder may have highest level of access possible and could delete the entire file system, or create a new account or password file resulting in incalculable damage. For example "rlogin" hold in the IBM RS-6000 workstations, which enabled a cracker to create a "root" shell or super user access mode.

(c) **Inconsistent Usage Threats :** Some times due to some specific combination of software and hardware, system becomes very open to security point of view.

We can use various protection method at various levels to overcome these security threats such as :

(1) **Trusted Base Security :** In which we trust everyone can do nothing extra for protection. But now a days this case is obselete.

(2) **Applying Password Schemes :** We can apply passwords at various level in clients and server computers to restrict unauthorized user to intract with servers as well as client.

(3) **Applying Biometric Passwords** may be the key to stop the unauthorize use of computers.

(4) **Applying Voice Matching Passwords** is the key now a days to authenticate the users.

Even we can restrict physical access through various approaches like biometric and voice matching etc.

9.2.5. Emerging Client Server Security Threats

Now, we've covered enough background information on client server that we can actually get into the security aspects of all of this. First of all, we'll get into the types of threats there are against networked computers, and then some things that can be done to protect ourself against various threats. Some main threats as follows :

1. Denial-of-Service

DoS (Denial-of-Service) attacks are probably the common, and most difficult to address. These are the common, because they're very easy to launch, difficult (sometimes impossible) to track, and it isn't easy to refuse the requests of the attacker, without also refusing legitimate requests for service.

Some things that can be done to reduce the risk of being stung by a denial of service attack include :

(a) Not running your visible-to-the-world servers at a level too close to capacity.

(b) Using packet filtering to prevent obviously forged packets from entering into your network address space.

(c) Keeping up-to-date on security-related patches for your hosts' operating systems.

2. Unauthorized Access

"Unauthorized access" is a very high-level term that can refer to a number of different sorts of attacks. The goal of these attacks is to access some resource that our machine should not provide the attacker. For example, a host might be a web server, and should provide anyone with requested web pages. However, that host should not provide command shell access without being sure that the person making such a request is someone who should get it, such as a local administrator.

3. Executing Commands Illegally

It's obviously undesirable for an unknown and untrusted person to be able to execute commands on your server machines. There are two main classifications of the severity of this problem: normal user access, and administrator access. A normal user can do a number of things on a system (such as read files, mail them to other people, etc.) that an attacker should not be able to do. This might, then, be all the access that an attacker needs. On the other hand, an attacker might wish to make configuration changes to a host (perhaps changing its IP address, putting a start-up script in place to cause the machine to shut down every time it's started, or something similar). In this case, the attacker will need to gain administrator privileges on the host.

4. Confidentiality Breaches

We need to examine the threat model: what is it that we are trying to protect ourself against ? There is certain information that could be quite damaging if it fell into the hands of a competitor, an enemy, or the public. In these cases, it's possible that compromise of a normal user's account on the machine can be enough to cause damage (perhaps in the form of public relation or obtaining information that can be used against the company, etc.).

5. Destructive Behavior

Among the destructive sorts of break-ins and attacks, there are two major categories.

(a) Data Diddling : The data diddler is likely the worst sort, since the fact of a break-in might not be immediately obvious. Perhaps he's toying with the numbers in your spreadsheets, or changing the dates in your projections and plans. Maybe he's changing the account numbers for the auto-deposit of certain paychecks. In any case, rare is the case when you'll come in to work one day, and simply know that something is wrong. An accounting procedure might turn up a discrepancy in the books three or four months after the fact. Trying to track the problem down will certainly be difficult, and once that problem is discovered, how can any of your numbers from that time period be trusted? How far back do you have to go before you think that your data is safe ?

(b) Data Destruction : Some of those perpetrate attacks are simply twisted jerks who like to delete things. In these cases, the impact on our computing capability—and consequently our business—can be nothing less than if a fire or other disaster caused our computing equipment to be completely destroyed.

9.3. FIREWALLS AND NETWORK SECURITY [MDU Dec. 2006]

Definition

Firewall is a software which enforces an access control policy between an internal network in an organization and the rest of the Internet. It may also be used for access control between two parts of the same internal network. A firewall blocks some Internet traffic and permits some other traffic.

Some firewalls are more restrictive than others. It's important to understand that a firewall implements a policy, so it's only as good as the policy it implements. If the policy is inconsistent or not well thought of, then a firewall may not be a good protection.

TIPS

The main thing firewalls protect against is unauthorized login. In addition they may block some or all of the outside traffic, while permitting all or most of the traffic from the inside to the outside. The traffic that usually is permitted is e-mail (both to and from the network) and HTTP access from the inside to the outside of the firewall. Protocols that may or may not be permitted are FTP, SSL, database connections, and so on. In addition to access control, firewalls keep statistics of the Internet packages and requests, and may be configured to issue a warning if some suspicious activity is going on.

There are two basic types of firewalls: *network layer* and *application layer* firewalls. Recently the distinction between the two is becoming somewhat blurry.

9.3.1. Importance of Firewall

The Internet, like any other society, is plagued with the kind of jerks who enjoy the electronic equivalent of writing on other people's walls with spray paint, tearing their mailboxes off, or just sitting in the street blowing their car horns. Some people try to get real work done over the Internet, and others have sensitive or proprietary data they must protect. Usually, **a firewall's purpose is to keep the jerks out of our network while still letting us get our job done.**

Many traditional-style corporations and data centers have computing security policies and practices that must be followed. In a case where a company's policies dictate how data must be protected, a firewall is very important, since it is the embodiment of the corporate policy. Frequently, the hardest part of hooking to the Internet, if we are a large company, is not justifying the expense or effort, but convincing management that it's safe to do so. **A firewall provides not only real security—it often plays an important role as a security blanket for management.**

A firewall can act as your corporate "ambassador" to the Internet. Many corporations use their firewall systems as a place to store public information about corporate products and services, files to download, bug-fixes, and so forth. Several of these systems have become important parts of the Internet service structure and have reflected well on their organizational sponsors. Note that while this is historically true, most organizations now place public information on a Web server, often protected by a firewall, but not normally on the firewall itself.

9.3.2. Firewall Protection

Some firewalls permit only email traffic through them, thereby protecting the network against any attacks other than attacks against the email service. Other firewalls provide less strict protections, and block services that are known to be problems.

Generally, firewalls are configured to protect against unauthenticated interactive logins from the "outside" world. This, more than anything, helps prevent intruders from logging into machines on your network. More elaborate firewalls block traffic from the outside to the inside, but permit users on the inside to communicate freely with the outside. The firewall can protect us against any type of network-borne attack if we unplug it.

TIPS

Firewalls are also important since they can provide a single "choke point" where security and audit can be imposed. Unlike in a situation where a computer system is being attacked by someone dialing in with a modem, the firewall can act as an effective "phone tap" and tracing tool. Firewalls provide an important logging and auditing function; often they provide summaries to the administrator about what kinds and amount of traffic passed through it, how many attempts there were to break into it, etc.

Because of this, firewall logs are critically important data. They can be used as evidence in a court of law in most countries. We should safeguard, analyze and protect our firewall logs accordingly.

9.3.3. Limitations of Firewall

Firewalls can't protect against attacks that don't go through the firewall. Many corporations that connect to the Internet are very concerned about proprietary data leaking out of the company through that route. *For a firewall to work, it must be a part of a consistent overall organizational security architecture.* Firewall policies must be realistic and reflect the level of security in the entire network. For example, a site with top secret or classified data doesn't need a firewall at all: they shouldn't be hooking up to the Internet in the first place, or the systems with the really secret data should be isolated from the rest of the corporate network.

Another thing a firewall can't really protect us against is traitors or idiots inside our network. While an industrial spy might export information through our firewall, he's just as likely to export it through a telephone, FAX machine, or Compact Disc. CDs are a far more likely means for information to leak from our organization than a firewall. Firewalls also cannot protect us against stupidity. Users who reveal sensitive information over the telephone are good targets for social engineering; an attacker may be able to break into our network by completely bypassing our firewall, if he can find a "helpful" employee inside who can be fooled into giving access to a modem pool.

Lastly, firewalls can't protect against bad things being allowed through them. For instance, many Trojan Horses use the Internet Relay Chat (IRC) protocol to allow an attacker to control a compromised internal host from a public IRC server. **If we allow any internal system to connect to any external system, then our firewall will provide no protection from this vector of attack.**

9.3.4. Basic Design Decisions in a Firewall

There are a number of basic design issues that should be addressed by the lucky person who has been tasked with the responsibility of designing, specifying, and implementing or overseeing the installation of a firewall.

The first and most important decision reflects the policy of how our company or organization wants to operate the system: is the firewall in place explicitly to deny all services except those critical to the mission of connecting to the Net, or is the firewall in place to provide a metered and audited method of "queuing" access in a non-threatening manner? There are degrees of paranoia between these positions; the final stance of your firewall might be more the result of a political than an engineering decision.

The second is: what level of monitoring, redundancy, and control do we want? Having established the acceptable risk level by resolving the first issue, you can form a checklist of what should be monitored, permitted, and denied.

The third issue is financial. We can't address this one here in anything but vague terms, but it's important to try to quantify any proposed solutions in terms of how much it will cost either to buy or to implement.

On the technical side, there are a couple of decisions to make, based on the fact that for all practical purposes what we are talking about is a static traffic routing service placed between the network service provider's router and your internal network. The traffic routing service may be implemented at an IP level via something like screening rules in a router, or at an application level via proxy gateways and services.

The decision to make is whether to place an exposed stripped-down machine on the outside network to run proxy services for telnet, FTP, news, etc., or whether to set up a screening router as a filter, permitting communication with one or more internal machines. There are pluses and minuses to both approaches, with the proxy machine providing a greater level of audit and potentially security in return for increased cost in configuration and a decrease in the level of service that may be provided (since a proxy needs to be developed for each desired service). The old trade-off between ease-of-use and security comes back to haunt us with a vengeance.

9.3.5. Selecting Firewalls

Firewall implementation and design for an enterprise can be a daunting task. Choices made early in the design process can have far-reaching security implications for years to come. In this series of tips, we take a detailed look at the task of firewall implementation.

We've divided the process into four stages:

1. **How to choose a firewall :** Despite the development and evolution of next generation security technologies, the firewall remains a vital component of any network architecture. However, today's organizations have several types of firewalls to choose from. This tip outlines five basic questions to ask when identifying the type of firewall that best suits our organization's network security needs.
2. **Choosing the right firewall topology :** When developing a perimeter protection strategy for an organization, one of the most common questions is "Where should I place firewalls for maximum effectiveness?" This tip provides an overview of the three most common firewall topologies, including diagrams of a bastion host, screened subnet and dual-firewall architectures.
3. **Placing systems in a firewall topology :** Once we have decided which topology best suits your IT infrastructure, it's time to decide where to place individual firewall systems within the chosen topology. This tip reviews several factors to consider when deciding where to place systems in a firewall topology, such as bastion host, screened subnet and multi-homed firewalls.
4. **Auditing firewall activity :** Firewall configurations change quickly and often, making it difficult to keep on top of routine firewall maintenance tasks. In this tip, learn ways to audit a firewall's capabilities to help keep network events in order.

When selecting firewalls, the following considerations should be taken into account :

- Ease of installation/configuration
- Does the firewall run without user intervention?
- Are there parameters that have to be set, and is it easy to do?

- Is there online help or technical support available?
- Does the firewall provide audit reports identifying time, location and type of attack?
- Is the cost of the firewall appropriate to the size of your business/office?
- Are maintenance/ monitoring requirements suitable for the size and type of business?
- What will be the training requirements for the firewall?
- Will the firewall have a significant impact on the operation of the system as a whole?

9.3.6. Types of Firewall

Firewalls have evolved into several distinct types. The following passages will help clarify the functions of the most common types of firewalls in order to help us understand which basic features that we may specifically require. Once we are educated in firewall types then we should feel more confidence in deciding whether to purchase *firewall software vs. hardware.* Common types of firewalls are as follows :

(i) Application Gateways of Proxy Gateways.
(ii) Packet Filtering.
(iii) Circuit Gateways.
(iv) Stateful Packet Inspection.
(v) Internet Connection Firewall.
(vi) Hybrid Systems.

9.3.6.1. Application Gateways or Proxy Gateways

The first firewalls were application gateways, and are sometimes known as proxy gateways. These are made up of hosts that run special software to act as a proxy server. This software runs at the *Application Layer* of our old friend the ISO/OSI Reference Model, hence the name. Clients behind the firewall must be *proxitized* (that is, must know how to use the proxy, and be configured to do so) in order to use Internet services. Traditionally, these have been the most secure, because they don't allow anything to pass by default, but need to have the programs written and turned on in order to begin passing traffic.

These are also typically the slowest, because more processes need to be started in order to have a request serviced. Figure 9.2 shows a application gateway.

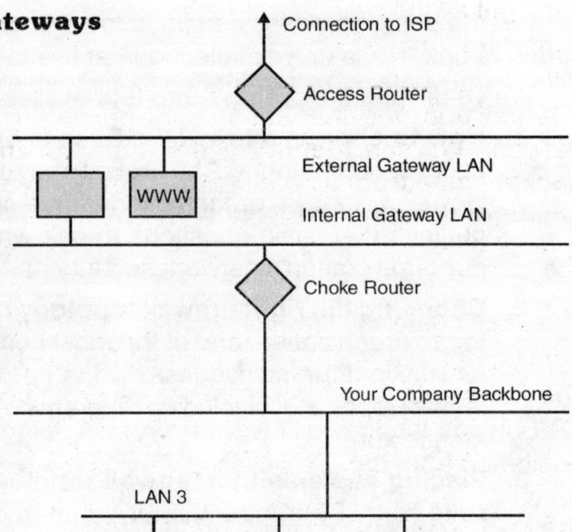

FIGURE 9.2. A Sample Application Gateway.

9.3.6.2. Packet Filtering

Packet filtering is a technique whereby routers have *ACLs* (Access Control Lists) turned on. By default, a router will pass all traffic sent in and will do so without any sort of restrictions. Employing ACLs is a method for enforcing our security policy with regard to what sorts of access we allow the outside world to have to our internal network, and vice versa.

TIPS

There is less overhead in packet filtering than with an application gateway, because the feature of access control is performed at a lower ISO/OSI layer (typically, the transport or session layer). Due to the lower overhead and the fact that packet filtering is done with routers, which are specialized computers optimized for tasks related to networking, a packet filtering gateway is often much faster than its application layer cousins.

Figure 9.3 shows a packet filtering gateway.

Because we're working at a lower level, supporting new applications either comes automatically, or is a simple matter of allowing a specific packet type to pass through the gateway.

There are problems with this method, though. Remember, TCP/IP has absolutely no means of guaranteeing that the source address is really what it claims to be. As a result, we have to use layers of packet filters in order to localize the traffic. We can't get all the way down to the actual host, but with two layers of packet filters, we can differentiate between a packet that came from the Internet and one that came from our internal network. **We can identify which network the packet came from with certainty, but we can't get more specific than that.**

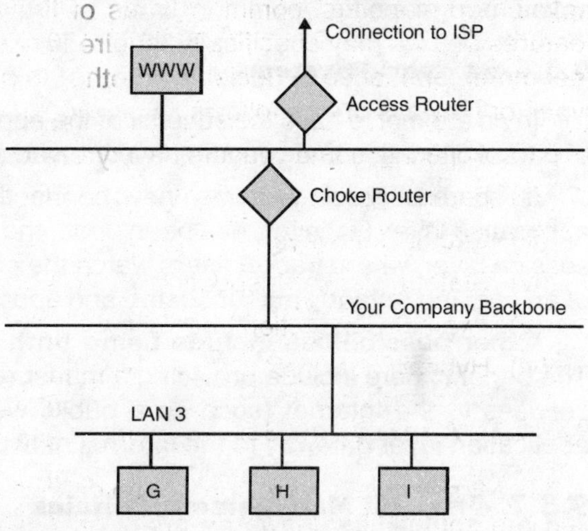

FIGURE 9.3. A Sample Packet Filtering Gateway.

9.3.6.3. Circuit Gateways

Circuit gateway firewalls work on the transport level of the protocol stack. They are fast and transparent, but really provide no protection from attacks. Circuit gateway firewalls also do not check the data in the packet. The one great benefit to this type of firewall is that they make the LAN behind the firewall invisible, as everything coming from within the firewall appears to have originated from the firewall itself. This is the least used type of firewall.

9.3.6.4. Stateful Packet Inspection

A fourth method that can be utilized by firewalls is called "Stateful Packet Inspection". It is called "Stateful" because it examines the contents of the packet to determine what the state of the communication is. It ensures that the stated destination computer has previously acknowledged the communication from the source computer. In this way all communications are initiated by the "receiving" computer and are taking place only with sources that are known or trusted from previous communication connections. In addition Stateful Packet Inspection firewalls are also more rigorous in their packet inspections. Stateful Packet Inspection firewalls also close off ports until an authorized connection is requested and acknowledged by the receiving computer. This allows for an added layer of protection from the threat of "port scanning" a method used by hackers to determine what PC services or applications are available to be utilized to gain access to the host computer.

9.3.6.5. Internet Connection Firewall

Windows XP provides Internet security in the form of the new Internet Connection Firewall (ICF). ICF makes use of active packet filtering, which means the ports on the firewall are opened for as long as needed to enable us to access the services we are interested in. The type of technology prevents hackers from scanning our computer's ports and resources. If we are hosting an Internet session, ICF allows us to open holes in the firewall that allow traffic on specific ports. This is called "port mapping."

9.3.6.6. Hybrid Systems

In an attempt to have the security of the application layer gateways with the flexibility and speed of packet filtering, some vendors have created systems that use the principles of both.

In some of these systems, new connections must be authenticated and approved at the application layer. Once this has been done, the remainder of the connection is passed down to the session layer, where packet filters watch the connection to ensure that only packets that are part of an ongoing (already authenticated and approved) conversation are being passed.

Other possibilities include using both packet filtering and application layer proxies. The benefits here include providing a measure of protection against your machines that provide services to the Internet (such as a public web server), as well as provide the security of an application layer gateway to the internal network.

9.3.7. Firewall Management Policies

With increased internet usage, escalating online attacks and unauthorised access, it is critical that businesses have systems in place to safeguard confidential data and network resources.

Firewalls are an integral part of the security strategy of leading companies, serving to defend the network edge, server layers, backend ingress points and to enforce 'End Point' security. It is vital that such a core component of network security is managed effectively. It can be a daunting task to determine the best firewall solution for your situation and you may be concerned about future developments in either the firewall market or your business needs. Running a firewall effectively is rarely easy. Setting and maintaining the rule base, interpreting and reacting appropriately to alerts, keeping software up-to-date, analysing log and audit files - all these tasks require time and skills that are often in short supply. If any of these tasks are not performed by experienced professionals, our network security is in danger.

TIPS

Security threats posed by IM (Instant Messaging), P2P (peer-to-peer) networking and other emerging communications modes are making firewall management a more important — and increasingly challenging — task.

Here's a checklist to help us determine if our organization's firewall-management policies are keeping pace with new threats.

1. **All network services are given a thorough security-risk analysis :** The reason for this is simple: to ensure that existing services, particularly emerging and unauthorized ones, will not create a security threat that the firewall is unable to handle.

2. **The firewall policy contains, or refers to, a policy that defines acceptable use from internal as well as external perspectives :** Firewall policies have to be carefully thought out and should reflect the views of technical and business executives, managers, and staff. This is especially important with technologies like IM, where defining legitimate use and misuse can be a touchy matter.
3. **The firewall policy does not conflict with established security and business policies :** This is another no-brainer. The firewall policy needs to reflect company policies — not create its own, possibly conflicting, rules.
4. **Firewall polices are regularly reviewed for relevance and potential gaps :** This concept is at the heart of protecting a network against emerging threats. If our firewall policies were developed a few years ago, they probably do not adequately reflect today's security threats. It is a good idea to review firewall policies at least every six months.
5. **Only firewall-system administrators can make changes :** Change implementation must be a centralized, coordinated effort, not an ad hoc exercise. To do otherwise invites confusion, conflicts and the likelihood that key security threats are left unaddressed.
6. **All firewall logs are regularly reviewed for security incidents :** This concept is simple, yet critical. You should know whether someone is tampering with our network.
7. **The firewall's default condition is to deny traffic :** This means that if an ACL (Access Control List) does not allow a specific type of traffic to pass the firewall, the firewall should deny the traffic.
8. **The firewall denies external traffic to itself :** This policy will help deflect attacks directed at the firewall.
9. **The firewall denies any traffic on the external interface that appears to come from an internal address :** This protocol will stifle spoofing, a situation in which a person or program successfully masquerades as another by falsifying data.
10. **Firewall monitoring is configured for real-time alerting :** If an incident arises, it is crucial that the network administrator is immediately notified of the condition so that he or she can take prompt action to address the situation.
11. **New technologies are periodically tested against the firewall to determine that the system is performing as designed :** When it comes to security, it pays to think like an attacker. Encourage organization's IT workers to launch supervised "attacks" against our firewall using the latest technologies and techniques.

SUMMARY

- Personal computers are often connected to a shared network. Personal computers in large companies are connected to large corporate networks. Personal computers in small companies are connected to a small local network, and computers in private homes often share a network between family members.
- When you are connected to the Internet, an IP address is used to identify your PC. If you don't protect yourself, this IP address can be used to access your computer from the outside world.
- When you are connected to the Internet, your shared resources can be accessed by the rest of the world.
- Client/server describes the relationship between two computer programs in which one program, the client, makes a service request from another program, the server, which fulfills the request.

- The main thing firewalls protect against is unauthorized login. In addition they may block some or all of the outside traffic, while permitting all or most of the traffic from the inside to the outside.

SHORT QUESTION ANSWERS

Q. 1. Define various Web security precautions?

Ans. The security policy should succinctly lay out your organization's policies with regard to:

- who is allowed to use the system
- when they are allowed to use it
- what they are allowed to do (different groups may be granted different levels of access)
- procedures for granting access to the system
- procedures for revoking access (e.g. when an employee leaves)
- what constitutes acceptable use of the system
- remote and local login methods
- system monitoring procedures
- protocols for responding to suspected security breaches

Q. 2. Define benefits of client server technology?

Ans. The following are what are considered to be the benefits of client/server computing.

- Many times easier to implement client/server than change a legacy application.
- New technology and the move to rapid application development such as object oriented technology.
- Long term cost benefits for development and support. Easy to add new hardware to support new systems such as document imaging and video teleconferencing which would not be feasible or cost efficient in a mainframe environment. Can implement multiple vendor software tools for each application.

Q. 3. What is Firewalls?

Ans. "Firewall is software which enforces an access control policy between an internal network in an organization and the rest of the Internet. It may also be used for access control between two parts of the same internal network. A firewall blocks some Internet traffic and permits some other traffic".

Q. 4. What are the basic types of firewalls?

Ans. Conceptually, there are three types of firewalls:

1. Network layer
2. Application layer
3. Hybrids

Q. 4. What do you understand by Denial-of-Service?

Ans. **DoS (Denial-of-Service)** attacks are probably the nastiest, and most difficult to address. These are the nastiest, because they're very easy to launch, difficult (sometimes impossible) to track, and it isn't easy to refuse the requests of the attacker, without also refusing legitimate requests for service.

Q. 5. What do you understand by unauthorized Access?

Ans. Unauthorized access" is a very high-level term that can refer to a number of different sorts of attacks. The goal of these attacks is to access some resource that your machine should not provide the attacker. For example, a host might be a web server, and should provide anyone with requested web pages.

Q. 6. What is the importance of Firewall?

Ans. The Internet, like any other society, is plagued with the kind of jerks who enjoy the electronic equivalent of writing on other people's walls with spray paint, tearing their mailboxes off, or just sitting in the street blowing their car horns. Some people try to get real work done over the Internet, and others have sensitive or proprietary data they must protect. Usually, a firewall's purpose is to keep the jerks out of your network while still letting you get your job done.

Q. 7. What is Packet Filtering technology of firewalls?

Ans. Packet filtering is a technique whereby routers have *ACLs* (Access Control Lists) turned on. By default, a router will pass all traffic sent it, and will do so without any sort of restrictions. Employing ACLs is a method for enforcing your security policy with regard to what sorts of access you allow the outside world to have to your internal network, and vice versa.

Q. 8. Define the Application Gateway firewall?

Ans. The first firewalls were application gateways, and are sometimes known as proxy gateways. These are made up of bastion hosts that run special software to act as a proxy server. This software runs at the *Application Layer* of our old friend the ISO/OSI Reference Model, hence the name. Clients behind the firewall must be *proxitized* (that is, must know how to use the proxy, and be configured to do so) in order to use Internet services. Traditionally, these have been the most secure, because they don't allow anything to pass by default, but need to have the programs written and turned on in order to begin passing traffic.

REVIEW QUESTIONS

1. What is the need of the firewall in computer network?
2. Explain different type of firewall with their advantages?
3. What is the role of client server technology in Internet application?
4. What are the various security issues can arise when you connect your PC to internet?
5. Differentiate application gateway firewall and packet filtering firewall?

❑❑❑

10

CHAPTER

Privacy and Encryption Schemes

INSIDE THIS CHAPTER

10.1. REQUIREMENTS FOR TRANSACTION SECURITY

There are following three major requirement of transaction security :

1. **Transaction Privacy :** Transactions must be kept private and secure in the sense that intruders cannot understand the message content.
2. **Transaction Confidentiality :** Confidentiality is important for transactions involving sensitive data such as credit card numbers, and will become even more important when data, such as employee records and social security numbers, begin traversing the network.

   The electronic commerce environment must ensure that all message traffic is confidential. After successful delivery to their destination gateways, messages must be removed (expunged) from the public environment, leaving only the accounting record of entry and delivery, including message length, authentication data, and perhaps the audit trail of message transfer agents.
3. **Transaction Integrity :** In order for electronic commerce to succeed, the contents of electronic commerce transactions must remain unmodified during transport between the client and the server. Transmission must be tamper-proof in the sense that no one can add, delete, or modify any part of the message during transit. Unauthorized combining of messages either by intermixing or concatenating during submission, validation, processing, or delivery should not be allowed.

10.2. INTERNET TRANSACTIONS AND SECURITY

The ease of shopping and comparing products and prices online has made it an attractive option for many shoppers. How can we make sure our transactions are safe and our credit card information going only where we intend it to ? There are several ways to help ensure safe transactions on the Internet, and more are becoming possible all the time. Some of these include :

1. Stored-value cards (cards that we can buy with specified, loaded Rs. amounts)
2. Smart cards (cards that can act as credit cards, debit cards and/or stored-value cards)
3. Point-of-sale (POS) devices (like our PDA or mobile phone)
4. Digital cash
5. E-wallets
6. Online payment services like PayPal

The most important method for paying for the things we purchase online is still the credit card. The following list provides some tips on how to make sure our transaction is secure and to do so following steps can be taken :

(a) **Use the Latest Internet Browser :** The program that we use to surf the Internet is called a browser. This software has built-in encryption capabilities that scramble the information we send to a server. Using the most recent browser ensures that the data is protected using the latest encryption technology. This technology also uses a Secure Sockets Layer (SSL), which is an Internet security protocol used by Internet browsers and Web servers to transmit sensitive information.

(b) **Look for Digital Certificates :** That authenticate the entity we are dealing with. Independent services like VeriSign will authenticate the identity of the Web site we are visiting. Web sites that use this service (usually those that sell items or services online) will have the VeriSign logo. By clicking on the logo, you can be assured that the site is legitimate, rather than a clone of the legitimate company set up to collect your personal and financial information.

(c) **Read the Privacy Policy :** The information we enter on the Web site should be kept confidential. Make sure we read the company's privacy policy to ensure that our personal information won't be sold to others.

(d) **Only Use One Credit Card** for all of your online purchases.

(e) Never give out passwords or user ID information online unless we know who we are dealing with **and why they need it.**

(f) **Keep Records of All of Our Internet Transactions.** Watch our credit card statement for the charges and make sure they're accurate.

(g) **After we have made Purchases Online, Check our E-mail.** Merchants often send confirmation e-mails or other communications about our order.

10.2.1. Types of Identity Theft

Identity theft can enter into many areas of our lives. It involves any instance where a person uses someone else's identification documents or other identifiers in order to impersonate that person for whatever reason. According to a September 2003 survey conducted by the Federal Trade Commission, an estimated 10 million people in the United States found out they were victims of identity theft in the previous year. More appropriately titled **identity fraud**, our identity might be stolen in order for someone to commit :

1. Financial Fraud : This type of identity theft includes bank fraud, credit card fraud, computer and telecommunications fraud, social program fraud, tax refund fraud, mail fraud, and several more.

2. Criminal Activities : This type of identity fraud involves taking on someone else's identity in order to commit a crime, enter a country, get special permits, hide one's own identity, or commit acts of terrorism. These criminal activities can include:

- Computer and cyber crimes
- Organized crime
- Drug trafficking
- Alien smuggling
- Money laundering

10.2.2. Credit Card Fraud

Have we ever eaten at a restaurant, paid with a credit card, and forgotten to get our copy of the credit card receipt ? Did we know that many of these receipts have our credit card number printed right there for anyone to see (and use) ? And, if we have signed them, our signature is also right there for someone to carefully copy. This can lead to the most simple form of identity theft. With this bit of information, some unscrupulous person can be well on his way to making purchases either by phone or on the Internet using our credit card number. We won't know about it until we get our statement (a good reason why we should always study the charges on our credit card statements!). All they have to have, in most cases, is our mailing address, which can be looked up in a phone book or easily found on the Internet.

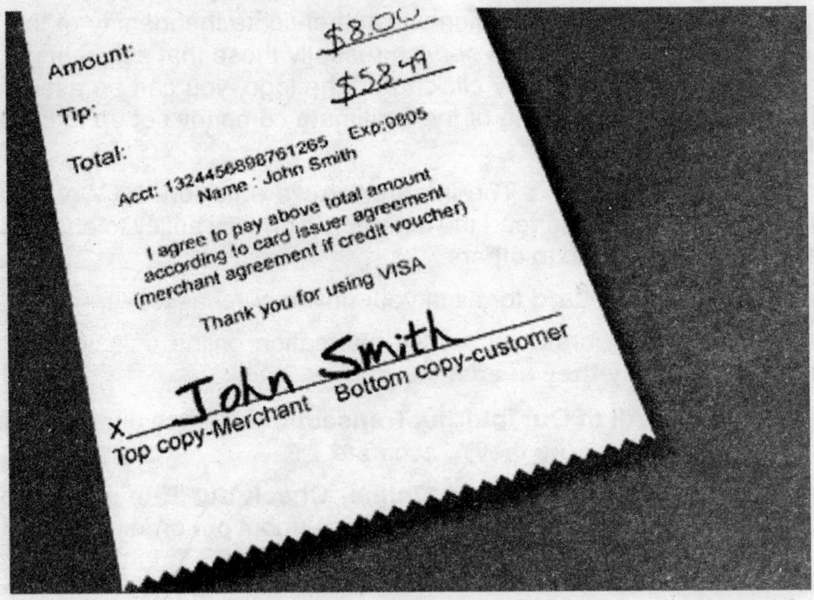

FIGURE 10.1.

Credit Card Fraud is identity theft in its most simple and common form. It can be accomplished either through a scenario like the one we just mentioned, or it can happen when our pre-approved credit card offers fall into the wrong hands. All a person has to do is get these out of our mailbox (or trash can) and mail them in with a change of address request and start spending. Someone can even apply for a credit card in our name if they have the right information. We won't know a thing about it until the credit card company tracks us down and demands payment for the purchases "we" have a racked up.

10.2.3. How To Protect Ourself

Protecting yourself from identity theft takes proactive effort. We can't simply assume it's not going to happen to us and go on about our life — it can happen to anyone. While you can't ever totally protect ourself from these thieves, you can at least make yourself less attractive as a victim by doing what you can to make it more difficult for them to access our information. Here are some things we can do to protect yourself :

1. **DON'T give out your ID number** unless it is absolutely necessary. Many companies collect more information than they really need. Make sure that it's something they have to have and make sure they'll protect your privacy.
2. **DESTROY any unwanted credit card offers.** This means rip, shred, burn, whatever you can do. These pre-approved offers come almost daily.
3. **DON'T put any other information besides your name and address on your checks,** and keep a close watch on your checkbook both when you're writing checks and when it is lying around. Someone can memorize your name, address and phone number during the short time it takes you to write a check.
4. **SHRED (cross-cut) any sensitive documents** before you throw them into the trash. This may seem like an extreme measure, but dumpster diving happens all the time and turns up a lot more personal information than you may realize.
5. **DON'T carry** your passport, or birth certificate in your wallet or purse. Also, only carry as many credit cards as are absolutely necessary. It has also been suggested that you photocopy everything you carry in your wallet to make canceling things easier in the event that your wallet is stolen.
6. **REVIEW your credit report** every year to make sure there haven't been any new credit cards or other accounts issued (to someone other than you) and to make sure there haven't been inquiries by people you haven't initiated business with.
7. **NEVER give out personal information** on the phone to someone you don't know and who initiated the call. Often, scam artists phone unsuspecting victims pretending to be their financial services company and request information to be provided over the phone. Usually, the story is to "update records" or sell a product. Get their name, phone number and address, and then call them back at the number you have on file or that is printed on the statements you receive.
8. **REVIEW your monthly credit card statement** each month to make sure there aren't any charges showing up that aren't yours. Also, make sure you *get* a monthly statement. If the statement is late, contact the credit card company. You never know when someone may have turned in a change-of-address form so they could make a few more weeks of purchases on your credit card without you noticing.
9. **DON'T mail bills or documents that contain personal data** (like tax forms or checks) from your personal mailbox. Take them directly to the post office or an official postal service mailbox.
10. **REACT QUICKLY if a creditor or merchant calls you about charges you didn't make.** This too may be the first notice you get that someone has stolen your identity. Get as much information from them as you can and investigate immediately.

11. **GUARD deposit slips as closely as you do checks.** Not only do they have your name, address and account number printed on them, but they can also be used to withdraw money from your account. All a thief has to do is write a bad check, deposit it into your account and use the "less cash received" line to withdraw your money.

10.3. ENCRYPTION

[MDU May 2005, 2007]

When we think of encryption, we probably think of some sort of secret code that prevents others from reading our messages. Whilst this privacy aspect of cryptography is important, it is only one of four aspects that are of particular importance in electronic commerce.

(i) **Authentication** allows customers to be sure that the merchant they are sending their credit card details to is who they say they are. It can also allow merchant's to verify that the customer is the real owner of the credit card being used.

(ii) **Integrity** ensures that the messages have not been tampered with by a third party during transmission.

(iii) **Non-repudiation** prevents customers or merchants denying they ever received or sent a particular message

(iv) **Privacy** prevents third parties from reading intercepted messages

10.3.1. Elements of an Encryption System

The main elements of an encryption system are the plaintext, the cryptographic algorithm, the key and the ciphertext.

(i) The **plaintext** is the raw message or data that is to be encrypted.

(ii) A **cryptographic algorithm or cipher** is a mathematical set of rules that defines how the plain text is to be combined with a key.

(iii) The **key** is a string of digits.

(iv) The **ciphertext** is the encrypted message.

These terms are probably best illustrated through a very simple example.

If we take the phrase "Web store" and add 2 characters to each letter the phrase becomes "ygd uvqtg". Here :

"Web store" is the plaintext

"add x characters to each letter" is the cryptographic algorithm

"2" is the key

"ygd uvqtg" is the ciphertext

FIGURE 10.2. Secret Key Encryption.

There are two main types of encryption in common use today : secret-key and public-key.

10.3.2. Secret-Key Encryption [MDU May 2007, 2008]

Secret-key encryption, also known as **single-key or symmetric encryption**, involves the use of a single key that is shared by both the sender and the receiver of the message.

After creating the message, the sender encrypts it with their key and passes it to the recipient who then decrypts it by using a copy of the same key used to encrypt it.

A widely used method of secret-key encryption is **the data encryption standard or DES.**

Secret-key encryption does have some limitations, particularly with regard to key distribution. Let us see some of them as follows :

(i) For privacy to be maintained, every transmitter of messages would need to provide a different key to everyone they intended to communicate with, otherwise every potential recipient would be able to read all messages whether it was intended for them or not.

(ii) While this is manageable where a small number of parties are involved (for example, sending a private e-mail to a friend) it is not practical for Web commerce which can involve communicating with thousands of customers.

(iii) Another limitation with secret-key encryption is its inability to support non-repudiation. As both parties share the same key it is possible for one party to create a message with the shared secret key and falsely claim the other party had sent it.

(iv) **Secret-key encryption on its own is not suitable for Web commerce—instead a system known as public-key encryption is used**

10.3.3. Public-Key Encryption [MDU May 2007, 2008]

Public-key encryption, or asymmetric encryption involves the use of two keys, one that can be used to encrypt messages (the public key) and one that can be used to either encrypt them or decrypt them (the private key).

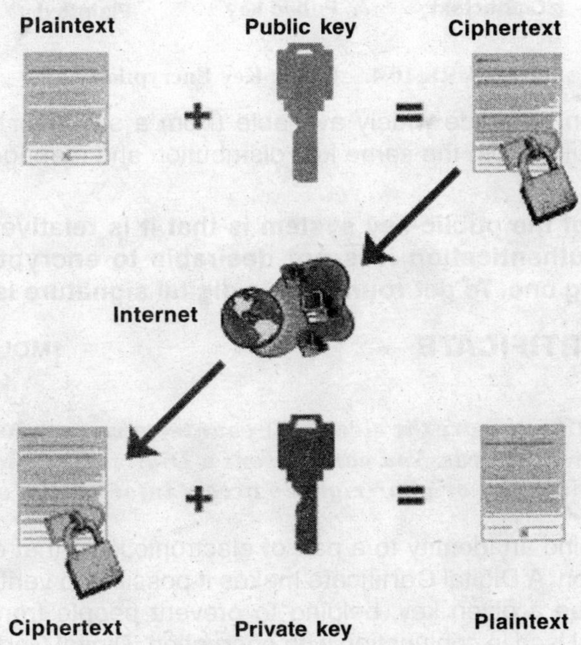

FIGURE 10.3.

These key pairs can be used in two different ways, to provide privacy or authentication.

Privacy is ensured by encoding a message with the public key as it can only be decoded by the holder of the private key.

TIPS

Authentication is achieved by encoding a message with the private key. Once the recipient has successfully decrypted it with the public key they can be assured it was sent by the holder of the private key.

FIGURE 10.4. **Public Key Encryption.**

As the public-key can be made widely available (from a server or third party), public-key cryptography does not suffer from the same key distribution and management problems as the secret-key system.

One disadvantage of the public-key system is that it is relatively slow, so when it is being used only for authentication it is not desirable to encrypt the whole message particularly if it is a long one. To get round this a digital signature is used.

10.4. DIGITAL CERTIFICATE **[MDU May 2006, Dec. 2007]**

Definition

Digital Certificates are the electronic counterparts to driver licenses, passports and membership cards. You can present a Digital Certificate electronically to prove your identity or your right to access information or services online.

Digital Certificates, bind an identity to a pair of electronic keys that can be used to encrypt and sign digital information. A Digital Certificate makes it possible to verify someone's claim that they have the right to use a given key, helping to prevent people from using phony keys to impersonate other users. Used in conjunction with encryption, Digital Certificates provide a more complete security solution, assuring the identity of all parties involved in a transaction.

A Digital Certificate is issued by a Certification Authority (CA) and signed with the CA's private key.

A Digital Certificate typically contains the :

- Owner's public key
- Owner's name
- Expiration date of the public key
- Name of the issuer (the CA that issued the Digital Certificate)
- Serial number of the Digital Certificate
- Digital signature of the issuer

The most widely accepted format for Digital Certificates is defined by the CCITT X.509 international standard; thus certificates can be read or written by any application complying with X.509. Further refinements are found in the PKCS standards and the PEM standard.

10.4.1. Use of Digital Certificates [MDU Dec. 2007]

Digital Certificates can be used for a variety of electronic transactions including e-mail, electronic commerce, groupware and electronic funds transfers. Netscape's popular Enterprise Server requires a Digital Certificate for each secure server.

For example, a customer shopping at an electronic mall run by Netscape's server software requests the Digital Certificate of the server to authenticate the identity of the mall operator and the content provided by the merchant. Without authenticating the server, the shopper should not trust the operator or merchant with sensitive information like a credit card number. The Digital Certificate is instrumental in establishing a secure channel for communicating any sensitive information back to the mall operator.

10.4.2. Need of Digital Certificate

Virtual malls, electronic banking, and other electronic services are becoming more common place, offering the convenience and flexibility of round-the-clock service direct from your home. However, your concerns about privacy and security might be preventing you from taking advantage of this new medium for your personal business. Encryption alone is not enough, as it provides no proof of the identity of the sender of the encrypted information. Without special safeguards, you risk being impersonated online. Digital Certificates address this problem, providing an electronic means of verifying someone's identity. Used in conjunction with encryption, Digital Certificates provide a more complete security solution, assuring the identity of all parties involved in a transaction.

Similarly, a secure server must have its own Digital Certificate to assure users that the server is run by the organisation it claims to be affiliated with and that the content provided is legitimate

10.4.3. Digital Signature

Definition

A digital signature functions for electronic documents like a handwritten signature does for printed documents. The signature is an unforgeable piece of data that asserts that a named person wrote or otherwise agreed to the document to which the signature is attached.

A digital signature actually provides a greater degree of security than a handwritten signature. The recipient of a digitally signed message can verify both that the message originated from the person whose signature is attached and that the message has not been altered either intentionally or accidentally since it was signed. Furthermore, secure digital signatures cannot be repudiated; the signer of a document cannot later disown it by claiming the signature was forged.

In other words, Digital Signatures enable "authentication" of digital messages, assuring the recipient of a digital message of both the identity of the sender and the integrity of the message. The requirement form the digital signatures are following :

(i) The signature must be a bit pattern that is dependent on the message being signed.

(ii) To prevent forgery and denial, the signature must use some information unique to the sender.

(iii) The digital signature must be easy to generate.

(iv) The storage of a copy of the digital copy must be simple.

(v) Forging the signature must be computationally infeasible i.e., either by constructing a fraudulent signature for a given message, or constructing a new message with an existing signature.

(vi) The signature must be easy to verify and recognize.

10.4.3.1. Use of Digital Signature for Authentication

Suppose Ramesh wants to send a signed message to Naresh. He creates a message digest by using a hash function on the message. The message digest serves as a "digital fingerprint" of the message; if any part of the message is modified, the hash function returns a different result. Ramesh then encrypts the message digest with her private key. This encrypted message digest is the digital signature for the message.

Ramesh sends both the message and the digital signature to Naresh. When Naresh receives them, he decrypts the signature using Ramesh's public key, thus revealing the message digest. To verify the message, he then hashes the message with the same hash function Ramesh used and compares the result to the message digest he received from Ramesh. If they are exactly equal, Naresh can be confident that the message did indeed come from Ramesh and has not changed since she signed it. If the message digests are not equal, the message either originated elsewhere or was altered after it was signed.

TIPS

Note that using a digital signature does not encrypt the message itself. If Ramesh wants to ensure the privacy of the message, he must also encrypt it using Naresh's public key. Then only Naresh can read the message by decrypting it with his private key.

It is not feasible for anyone to either find a message that hashes to a given value or to find two messages that hash to the same value. If either were feasible, an intruder could attach a false message onto Ramesh's signature. Specific hash functions have been designed to have the property that finding a match is not feasible, and are therefore considered suitable for use in cryptography.

One or more Digital Certificates can accompany a digital signature. If a Digital Certificate is present, the recipient (or a third party) can check the authenticity of the public key.

10.4.3.2. Validity of Digital Signatures

Normally, a key expires after some period of time, such as one year, and a document signed with an expired key should not be accepted. However, there are many cases where it is necessary for signed documents to be regarded as legally valid for much longer than two years; long-term leases and contracts are examples. By registering the contract with a digital time-stamping service at the time it is signed, the signature can be validated even after the key expires.

If all parties to the contract keep a copy of the time-stamp, each can prove that the contract was signed with valid keys. In fact, the time-stamp can prove the validity of a contract even if one signer's key gets compromised at some point after the contract was signed. Any digitally signed document can be time-stamped, assuring that the validity of the signature can be verified after the key expires.

10.4.3.3. Digital Time-Stamping Service

A Digital Time-Stamping service (DTS) issues time-stamps which associate a date and time with a digital document in a cryptographically strong way. The digital time-stamp can be used at a later date to prove that an electronic document existed at the time stated on its time-stamp. For example, a physicist who has a brilliant idea can write about it with a word processor and have the document time-stamped. The time-stamp and document together can later prove that the scientist deserves the Nobel Prize, even though an arch rival may have been the first to publish.

The use of a DTS would appear to be extremely important, if not essential, for maintaining the validity of documents over many years. Suppose a landlord and tenant sign a twenty-year lease. The public keys used to sign the lease are set to expire after two years. Solutions such as recertifying the keys or resigning every two years with new keys require the cooperation of both parties several years after the original signing. If one party becomes dissatisfied with the lease, he or she may refuse to cooperate. The solution is to register the lease with the DTS at the time of the original signing; both parties would then receive a copy of the time-stamp, which can be used years later to enforce the integrity of the original lease.

In the future, it is likely that a DTS will be used for everything from long-term corporate contracts to personal diaries and letters. Today, if an historian discovers some lost letters of Gandhi Ji, their authenticity is checked by physical means. But a similar find 100 years from now may consist of an author's computer files; digital time-stamps may be the only way to authenticate the find.

10.5. SECURE SOCKET LAYER (SSL)

Definition

Secure Socket Layer (SSL) is a protocol designed and implemented by Netscape Communications. Netscape claims it is designed to work, as the name implies. At the socket layer, to protect any higher level protocol build on sockets, such as Telnet, FTP and HTTP.

Secure Socket Layer (SSL) technology allows web browsers and web servers to communicate over a secure connection. In this secure connection, the data that is being sent is encrypted before being sent and then is decrypted upon receipt and before processing. Both the browser and the server encrypt all traffic before sending any data. SSL addresses the following important security considerations.

(i) **Authentication :** During your initial attempt to communicate with a web server over a secure connection, that server will present our web browser with a set of credentials in the form of a server certificate. The purpose of the certificate is to verify that the site is who and what it claims to be. In some cases, the server may request a certificate that the client is who and what it claims to be (which is known as client authentication).

(ii) **Confidentiality :** When data is being passed between the client and the server on a network, third parties can view and intercept this data. SSL responses are encrypted so that the data cannot be deciphered by the third party and the data remains confidential.

(iii) **Integrity :** When data is being passed between the client and the server on a network, third parties can view and intercept this data. SSL helps guarantee that the data will not be modified in transit by that third party.

10.5.1. Use of SSL

To use SSL, an application server must have an associated certificate for each external interface, or IP address, that accepts secure connections. The theory behind this design is that a server should provide some kind of reasonable assurance that its owner is who you think it is, particularly before receiving any sensitive information. It may be useful to think of a certificate as a "digital driver's license" for an Internet address. It states with which company the site is associated, along with some basic contact information about the site owner or administrator.

The digital certificate is cryptographically signed by its owner and is difficult for anyone else to forge. For sites involved in e-commerce or in any other business transaction in which authentication of identity is important, a certificate can be purchased from a well-known Certificate Authority (CA).

Sometimes authentication is not really a concern—for example, an administrator may simply want to ensure that data being transmitted and received by the server is private and cannot be snooped by anyone eavesdropping on the connection. In such cases, you can save the time and expense involved in obtaining a CA certificate and simply use a self-signed certificate.

SSL uses *public key cryptography*, which is based on *key pairs*. Key pairs contain one public key and one private key. If data is encrypted with one key, it can be decrypted only with the other key of the pair. This property is fundamental to establishing trust and privacy in transactions. For example, using SSL, the server computes a value and encrypts the value using its private key. The encrypted value is called a *digital signature*. The client decrypts the encrypted value using the server's public key and compares the value to its own computed value. If the two values match, the client can trust that the signature is authentic, because only the private key could have been used to produce such a signature.

Digital certificates are used with the HTTPS protocol to authenticate web clients. The HTTPS service of most web servers will not run unless a digital certificate has been installed. Use the procedure outlined later to set up a digital certificate that can be used by your web server to enable SSL.

10.6. AUTHENTICATION [MDU May 2006]

As stated earlier, encryption is the process of taking all of the data that one computer is sending to another and encoding it into a form that only the other computer will be able to decode. Another process, **authentication**, is used to verify that the information comes from a trusted source. Basically, if information is "authentic," we know who created it and we know that it has not been altered in any way since that person created it. These two processes, encryption and authentication, work hand-in-hand to create a secure environment.

There are several ways to authenticate a person or information on a computer:

1. Password : The use of a user name and password provides the most common form of authentication. You enter your name and password when prompted by the computer. It checks the pair against a secure file to confirm. If either the name or the password does not match, then you are not allowed further access.

2. Pass Cards : These cards can range from a simple card with a magnetic strip, similar to a credit card, to sophisticated smart cards that have an embedded computer chip.

3. Digital Signatures : A digital signature is basically a way to ensure that an electronic document (e-mail, spreadsheet, text file) is authentic.

Recently, more sophisticated forms of authentication have begun to show up on home and office computer systems. Most of these new systems use some form of **biometrics** for authentication. Biometrics uses biological information to verify identity. Biometric authentication methods include:

(a) Fingerprint scan

(b) Retina scan

(c) Face scan

(d) Voice identification

10.7. INTRODUCTION TO VPN

Definition

A Virtual Private Network (VPN) connects the components of one network over another network. VPNs accomplish this by allowing the user to tunnel through the Internet or another public network in a manner that provides the same security and features formerly available only in private networks.

VPNs allow users working at home or on the road to connect in a secure fashion to a remote corporate server using the routing infrastructure provided by a public internetwork (such as the Internet). From the user's perspective, the VPN is a point-to-point connection between the user's computer and a corporate server. The nature of the intermediate internetwork is irrelevant to the user because it appears as if the data is being sent over a dedicated private link.

VPN technology also allows a corporation to connect to branch offices or to other companies over a public internetwork (such as the Internet), while maintaining secure communications. The VPN connection across the Internet logically operates as a Wide Area Network (WAN) link between the sites.

In both of these cases, the secure connection across the internetwork appears to the user as a private network communication—despite the fact that this communication occurs over a public internetwork—hence the name *Virtual Private Network*.

VPN technology is designed to address issues surrounding the current business trend toward increased telecommuting and widely distributed global operations, where workers must be able to connect to central resources and must be able to communicate with each other.

10.7.1. Common Uses of VPNs

Let us see VPN situations in more detail.

Remote User Access Over the Internet

VPNs provide remote access to corporate resources over the public Internet, while maintaining privacy of information. Figure 10.5 shows a VPN used to connect a remote user to a corporate intranet.

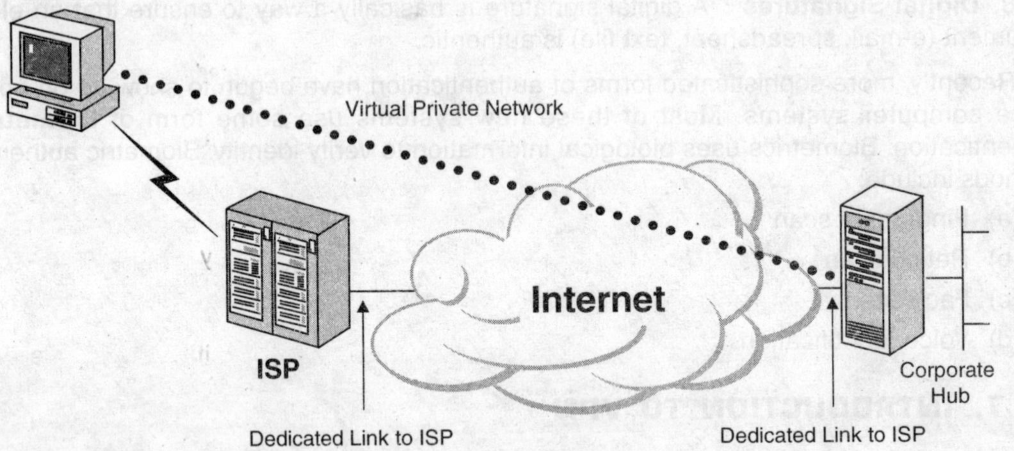

FIGURE 10.5. Using a VPN to connect a remote client to a private LAN.

Rather than making a long distance (or 1-800) call to a corporate or outsourced Network Access Server (NAS), the user calls a local ISP. Using the connection to the local ISP, the VPN software creates a virtual private network between the dial-up user and the corporate VPN server across the Internet.

Connecting Networks Over the Internet

There are two methods for using VPNs to connect local area networks at remote sites:

Using dedicated lines to connect a branch office to a corporate LAN. Rather than using an expensive long-haul dedicated circuit between the branch office and the corporate hub, both the branch office and the corporate hub routers can use a local dedicated circuit and local ISP to connect to the Internet. The VPN software uses the local ISP connections and the Internet to create a virtual private network between the branch office router and corporate hub router.

Using a dial-up line to connect a branch office to a corporate LAN. Rather than having a router at the branch office make a long distance (or 1-800) call to a corporate or outsourced NAS, the router at the branch office can call the local ISP. The VPN software uses the connection to the local ISP to create a VPN between the branch office router and the corporate hub router across the Internet.

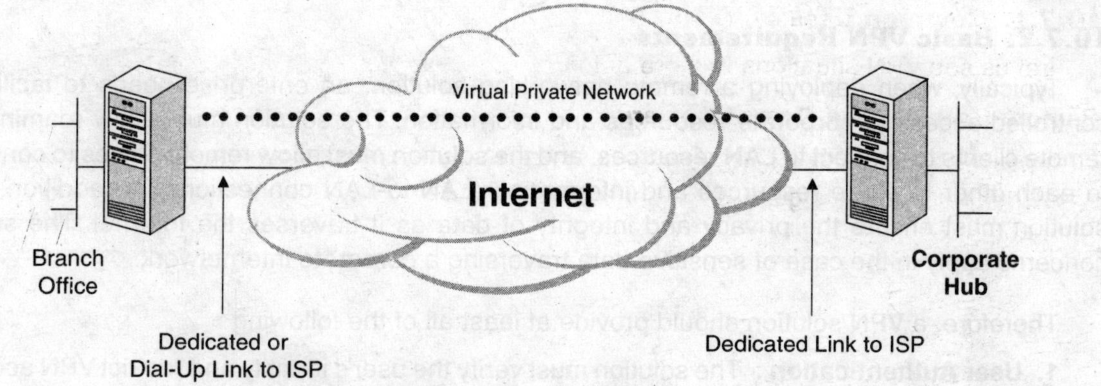

FIGURE 10.6. Using a VPN to Connect Two Remote Sites.

In both cases, the facilities that connect the branch office and corporate offices to the Internet are local. The corporate hub router that acts as a VPN server must be connected to a local ISP with a dedicated line. This VPN server must be listening 24 hours a day for incoming VPN traffic.

Connecting Computers Over an Intranet

In some corporate internetworks, the departmental data is so sensitive that the department's LAN is physically disconnected from the rest of the corporate internetwork. Although this protects the department's confidential information, it creates information accessibility problems for those users not physically connected to the separate LAN.

VPNs allow the department's LAN to be physically connected to the corporate internetwork but separated by a VPN server. The VPN server is not acting as a router between the corporate internetwork and the department LAN. A router would connect the two networks, allowing everyone access to the sensitive LAN. By using a VPN, the network administrator can ensure that only those users on the corporate internetwork who have appropriate credentials (based on a need-to-know policy within the company) can establish a VPN with the VPN server and gain access to the protected resources of the department. Additionally, all communication across the VPN can be encrypted for data confidentiality. Those users who do not have the proper credentials cannot view the department LAN.

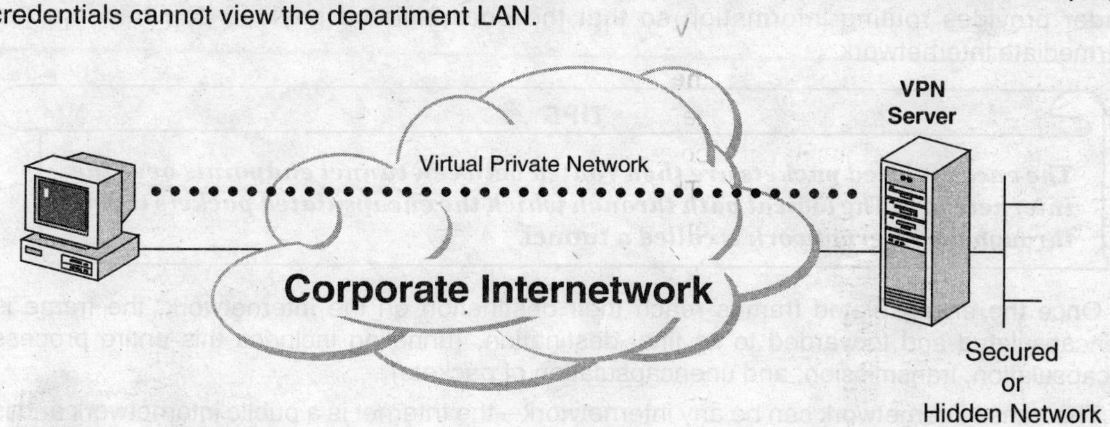

FIGURE 10.7. Using a VPN to Connect Two Computers on the Same LAN.

10.7.2. Basic VPN Requirements

Typically, when deploying a remote networking solution, an enterprise needs to facilitate controlled access to corporate resources and information. The solution must allow roaming or remote clients to connect to LAN resources, and the solution must allow remote offices to connect to each other to share resources and information (LAN-to-LAN connections). In addition, the solution must ensure the privacy and integrity of data as it traverses the Internet. The same concerns apply in the case of sensitive data traversing a corporate internetwork.

Therefore, a VPN solution should provide at least all of the following :

1. **User Authentication** : The solution must verify the user's identity and restrict VPN access to authorized users only. It must also provide audit and accounting records to show who accessed what information and when.

2. **Address Management** : The solution must assign a client's address on the private net and ensure that private addresses are kept private.

3. **Data Encryption** : Data carried on the public network must be rendered unreadable to unauthorized clients on the network.

4. **Key Management** : The solution must generate and refresh encryption keys for the client and the server.

5. **Multiprotocol Support** : The solution must handle common protocols used in the public network. These include IP, Internet Packet Exchange (IPX), and so on.

10.7.3. Tunneling Basics

Tunneling is a method of using an internetwork infrastructure to transfer data for one network over another network. The data to be transferred (or *payload*) can be the frames (or packets) of another protocol. Instead of sending a frame as it is produced by the originating node, the tunneling protocol encapsulates the frame in an additional header. The additional header provides routing information so that the encapsulated payload can traverse the intermediate internetwork.

TIPS

The encapsulated packets are then routed between tunnel endpoints over the internetwork. The logical path through which the encapsulated packets travel through the internetwork is called a tunnel.

Once the encapsulated frames reach their destination on the internetwork, the frame is unencapsulated and forwarded to its final destination. Tunneling includes this entire process (encapsulation, transmission, and unencapsulation of packets).

The transit internetwork can be any internetwork—the Internet is a public internetwork and is the most widely known real world example. There are many examples of tunnels that are carried over corporate internetworks.

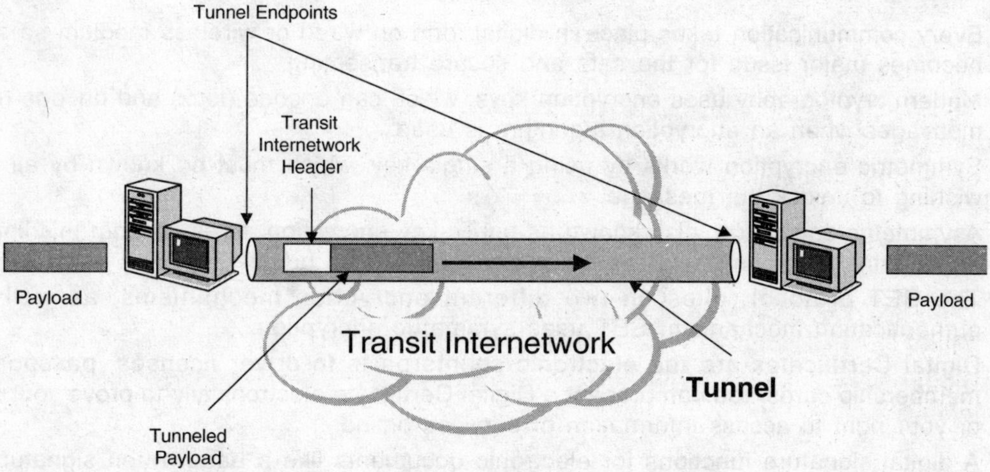

FIGURE 10.8. **Tunneling.**

10.7.4. Security Mechanisms are Available Through VPN

Following security mechanisms are available through VPN :

1. Authorisation : VPN connections are only created for users and routers that have been authorised. For Windows 2000, authorization of VPN connections is determined by dial-in properties on the user account and remote access policies. If a user or router is not authorised for such connections, the server will disable them.

2. Authentication : This is a vital security concern. Authentication takes place at two levels:

(*a*) **Machine-Level Authentication** : When IPSec protocol is used for a VPN connection, machine-level authentication is performed through the exchange of machine certificates during the establishment of the IPSec connection.

(*b*) **User-Level Authentication :** Before data can be sent over the PPTP (Point to Point Transfer Protocol) or L2TP (Layer 2 Tunneling Protocol) tunnel, the user must be authenticated. This is done through the use of a PPP authentication method.

3. Data Encryption – the protocols used to create VPN connections allow encrypted data to be sent over a network. Although it is possible to have a non-encrypted connection, this is not recommended. Note that data encryption for VPN connections does not provide end-to-end security (encryption), but only security between the client and the VPN server. In order to provide a secure end-to-end connection, the IPSec protocol can be used once a VPN connection has been established.

4. Packet Filtering – in order to enhance security of the VPN server, packet filtering must be configured so that the server only performs VPN routing. To this end, appropriate RRAS filters should be used (for Windows 2000) on the Internet interface of the VPN.

SUMMARY

- Every communication takes place in digital form on wired or wireless medium so security becomes major issue for the safe and secure transecting.
- Modern cryptography uses encryption keys, which can encode (lock) and decode (unlock) messages when an encryption algorithm is used.
- Symmetric encryption works by using a single key, which must be known by all parties wishing to unlock the message.
- Asymmetric encryption, also known as public-key encryption, frees us from this limitation. Asymmetric algorithms use two keys – a public and a private key.
- The SET protocol relies on two different encryption mechanisms, as well as an authentication mechanism. SET uses symmetric encryption.
- Digital Certificates are the electronic counterparts to driver licenses, passports and membership cards. You can present a Digital Certificate electronically to prove your identity or your right to access information or services online.
- A digital signature functions for electronic documents like a handwritten signature does for printed documents.

SHORT QUESTION ANSWERS

Q. 1. What is SET?

Ans. The SET protocol relies on two different encryption mechanisms, as well as an authentication mechanism. SET uses symmetric encryption, in the form of the aging Data Encryption Standard (DES), as well as asymmetric, or public-key, encryption to transmit session keys for DES transactions (IBM, 1998).

Q. 2. What are main elements of an encryption system?

Ans. The components are the plaintext, the cryptographic algorithm, the key and the cipher text.

- The **plaintext** is the raw message or data that is to be encrypted.
- A **cryptographic algorithm or cipher** is a mathematical set of rules that defines how the plain text is to be combined with a key.
- The **key** is a string of digits.
- The **ciphertext** is the encrypted message.

Q. 3. What is digital signature?

Ans. Digital Signatures enable "authentication" of digital messages, assuring the recipient of a digital message of both the identity of the sender and the integrity of the message.

Q. 4. What is SSL?

Ans. SSL, Secure Socket Layer, is a protocol designed and implemented by Netscape Communications. Netscape claims it is designed to work, as the name implies. At the socket layer, to protect any higher level protocol builds on sockets, such as Telnet, FTP and HTTP.

Q. 5. What is Public-key encryption?

Ans. Public-key encryption, or asymmetric encryption involves the use of two keys, one that can be used to encrypt messages (the public key) and one that can be used to either encrypt them or decrypt them (the private key).

Q. 6. Secret-key encryption

Ans. Secret-key encryption, also known as single-key or symmetric encryption, involves the use of a single key that is shared by both the sender and the receiver of the message. After creating the message, the sender encrypts it with their key and passes it to the recipient who then decrypts it by using a copy of the same key used to encrypt it.

REVIEW QUESTIONS

1. "Web security is major concern." Justify it and also discuss the some primary security services categories.
2. Now a days most of the e-commerce transactions takes place on internet. What steps we can take to ensure internet transaction security.
3. What do you understand by Encryption and what are various elements of Encryption system ?
4. Differentiate between Secret-key encryption and public-key encryption ?
5. What is the use of digital certification e-commerce ?
6. Define SSL and how it becomes functional.
7. "VPN can fulfill the need of corporate organization about secure treansmission." Justify the statement.

❑❑❑

APPENDICES

- Appendix A : Internet Lab
- Appendix B : Glossary

A

APPENDIX

INTERNET LAB

Practical Based on Internet Fundamentals Syllabus

Q. 1. How can you open a new mail account ?

Ans. We can open the new mail account in following steps :

(a) First select any free e-mail service provider like Rediffmail, yahoomail or Gmail. Let us assume we have selected Gmail.

(b) Go to www.mail.google.com to open your new Gmail account.

On the Create an Account page, start by filling in the asked information.

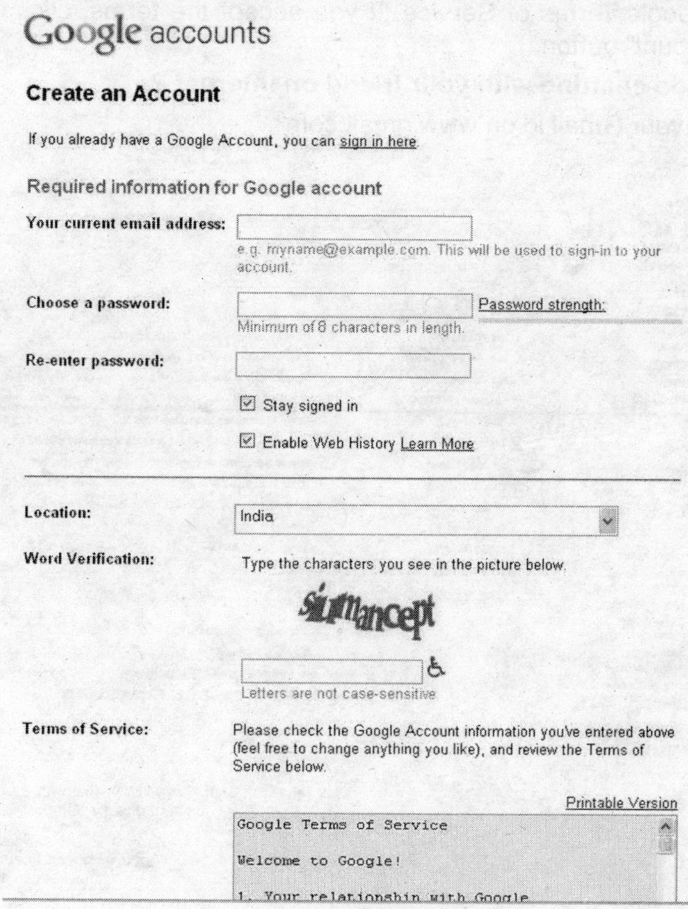

(a) Your first name, Your last name, Your desired login name (be sure to click on the check availability box to ensure you get the login name you want.)

(b) Select an 8 character minimum password. There is a password strength indicator that can let you know how strong your password is. This is an optional step that you can take, but it will ensure that your password would not be easy to "crack".

(c) Once you have decided on a password, you can check the "remember me on this computer" if you want Gmail to automatically insert your password once you have entered your login name. You can also check the "Enable Web History" box if you want Google to keep track of the sites that you visit most often.

(d) Enter a security question from the drop down menu. Try to choose one that would not be easy for someone else to answer.

(e) Type in the answer to your security question.

(f) If you choose, you can type in a secondary email account that will authenticate your Gmail account if there are ever any problems or if you forget your password. This is an optional step and can be skipped over if you do not have a secondary email or do not wish to enter it.

(g) Select your Country location from the drop-down menu.

(h) Type in the characters that you see for the word verification.

(i) Review the Google Terms of Service. If you accept the terms, click on the "I accept. Create my account" button.

Q. 2. How can you do chatting with your friend on internet ?

Ans. (a) Login with your Gmail id on www.gmail.com

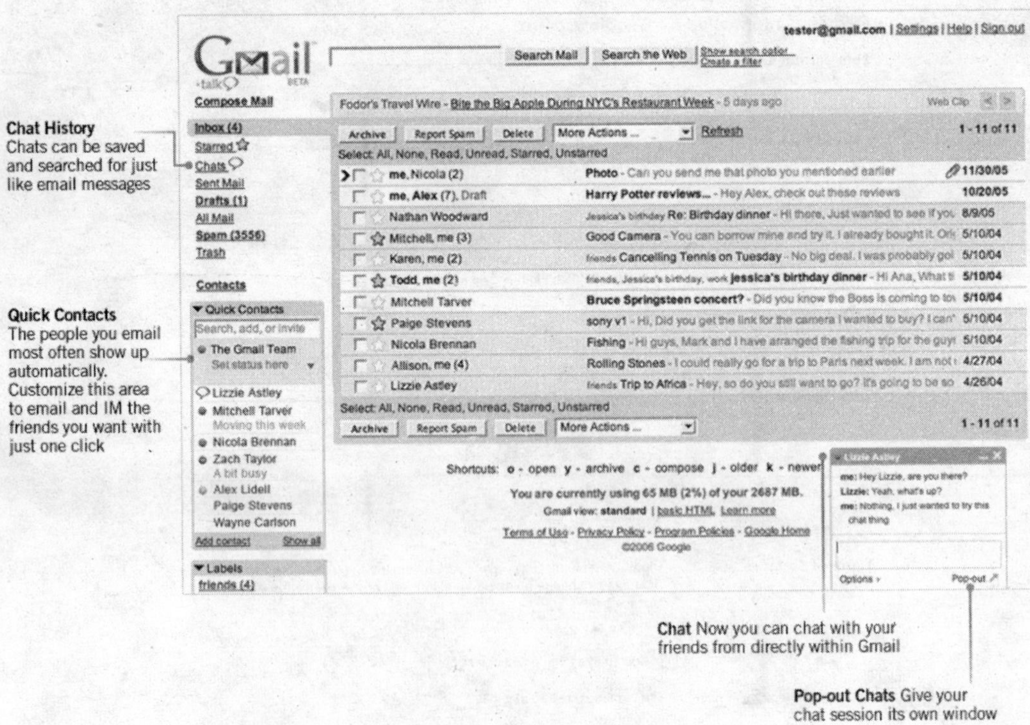

(b) You will find chat box on left hand side.

(c) On the top of chat box you can find text box.

(d) Enter your friends Gmail id in that text box and press enter (its invitation process).

(e) Tell your friend to Accept the invitation.

(f) If your friend is online you can find your friend name with green icon in chat box.

(g) Just click on your friends id and one chat box appears on right side.

(h) Start chatting with your friend.

3. What is Google earth? Try to use it.

Ans. Google Earth is a application software which provides map of the world . You can zoom and glide over stitched together satellite photos of the world. Use Google Earth to find driving directions, find nearby restaurants, measure the distance between two locations, do serious research, or go on virtual vacations. You can download Google Earth from http://earth.google.com/

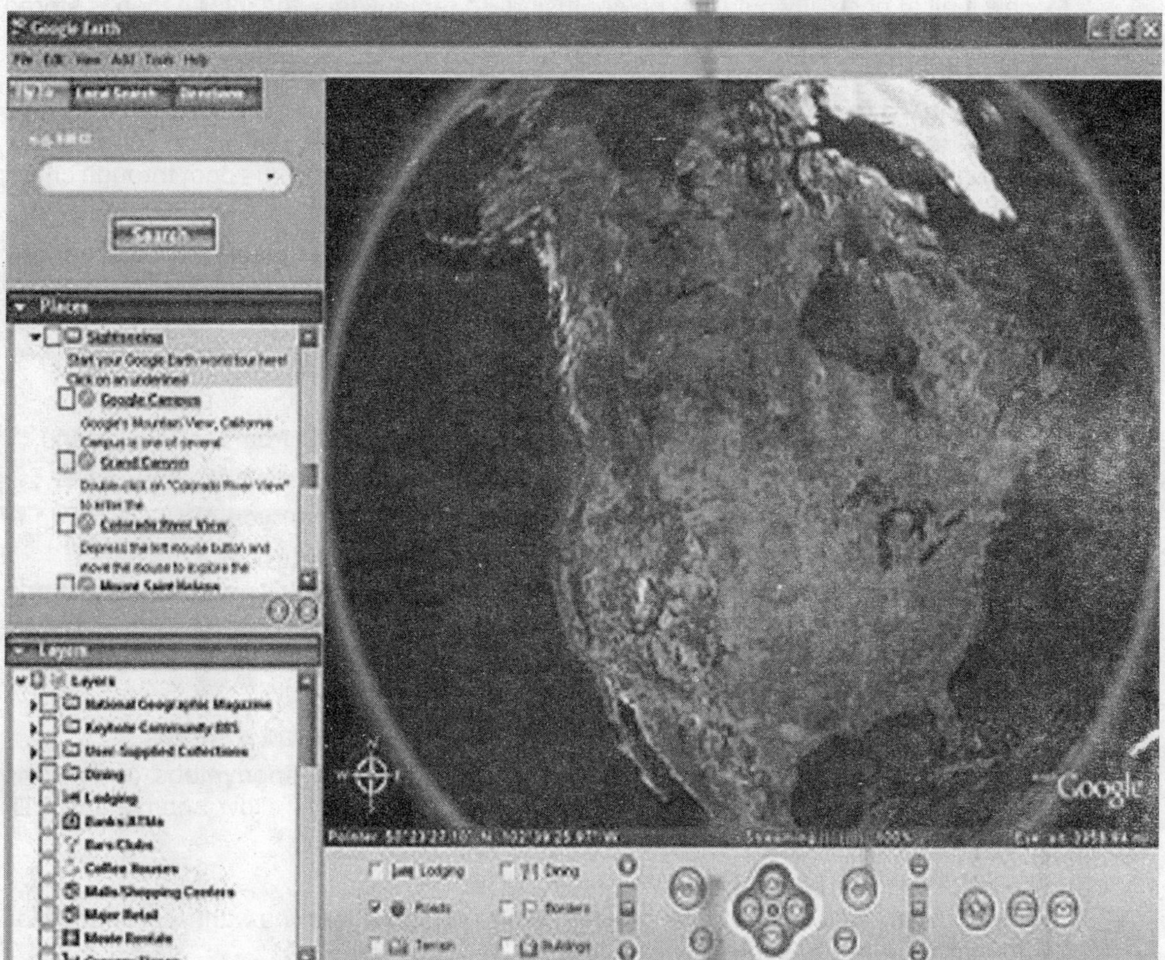

Most users won't require more than the free version.

The Google Earth Interface : Google Earth opens with a view of the world from space. Clicking and dragging on the planet will gently spin the globe. The middle scroll wheel or right-click dragging will zoom in and out for close-up views. In some areas the close-ups are detailed enough to make out cars and even people.

If you pass over the upper right-hand corner of the globe, the small compass will turn into the larger navigation control. Click and drag the circle to turn the map. North on the compass will move accordingly. Click on the arrows to move left or right, or use the star in the middle as a joystick to move in any direction. The dial to the right controls zoom levels.

Tilted View : You can tilt the globe to have a perspective view and move the horizon line up or down. This lets you view close-ups as if you were just above them, rather than viewing straight down. It also comes in very handy with the 3-D Buildings. This view is best with the Terrain layer turned on.

Layers : Google Earth can provide a lot of information about a location, and if you were to view it all at once, it would just be confusing. To remedy this, the information is stored in layers, which can be turned on or off. Layers include roads, border labels, parks, food, gas, and lodging.

Terrain and 3D Buildings : Two layers are useful for creating a more three dimensional globe. Terrain simulates the elevation levels, so when you tilt your view, you can see mountains and other terrain objects. The 3D Buildings layer lets you zoom through cities.

4. How can you use FTP on windows platform?

Ans. To use FTP you will need to install FTP software on your computer. There are many freely available FTP programs on the Internet such as, FTP explorer (http://www.ftpx.com) and Smart FTP (http://www.smartftp.com).

Follow the steps below to connect to a server, to upload files using FTP Explorer :

- Open FTP Explorer.
- Click on the connect button on the toolbar or select connect from the Tools menu.
- In the Profile Name text, enter a name to give your FTP connection.
- In the Host Address text box enter the address of the server you are connecting to (this can either be the DNS entry or the IP Address).
- Enter the Port number that the server accepts FTP connections on (the default port for FTP is 21).
- Tick the Use PASV option, if your computer is behind a firewall, which does not allow external FTP connections.
- Tick the Use Firewall option if your computer is located behind a firewall.
- Either enter a user name for the FTP login or tick the Anonymous option (the anonymous option will only work if the server is set up to allow anonymous FTP connections – if this is selected go to step 10).
- Enter the user's password in the Password text box.
- In the Initial path text box, enter the path on the server, where you want your connection to start at – this is optional.
- In the Attempts text box enter the number of times to try the FTP connection if fails, you will need to have at least 1 in this text box.

- In the Download path you can specify the path on your computer where to save the files downloaded from the server – this is optional, you can specify this later if you like.
- To establish the connection with the server click on the Connect button.

5. How do you upload files using FTP on windows platform?

Ans. Once you have established your connection with the server the next step is to upload your files to the server. The following describes how to do this :

- Go to the path on the server where you want to upload your files to.
- In the Tools menu click on the Upload button.
- Select the file or files that you want to upload (to select more than one file hold down the CTRL key).

  You can also upload files by dragging and dropping them to the directory/folder on the server.
- Click on the Open button. Your file will now be uploaded to the server. (You will see this uploading process in the dialog box.)

6. How do you download files using FTP on windows platform?

Ans. You can also use FTP to download files to your computer from a server. The following describes how to download a file.

- Make sure you are connected to the server.
- Go to the file that you want to download.
- Right click on the file and select Download To
- Select the path on your computer where you want the file to be saved to.
- The file will now be downloaded to your computer. (You will see this downloading process happening in the dialog box.)

7. How do you end your connection on windows platform?

Ans. You can end the connection by simply clicking on the Disconnect button on the tool bar or by clicking on Disconnect from the Tools menu.

8. How can you use Telnet server?

Ans. We will assume you're running Windows 98, but the procedure is similar for any type of computer or operating system.

First, you will need to open the Telnet program on your computer. Click on the **Start** button and select **Run....** In the dialog box that pops up, type telnet and click **OK**:

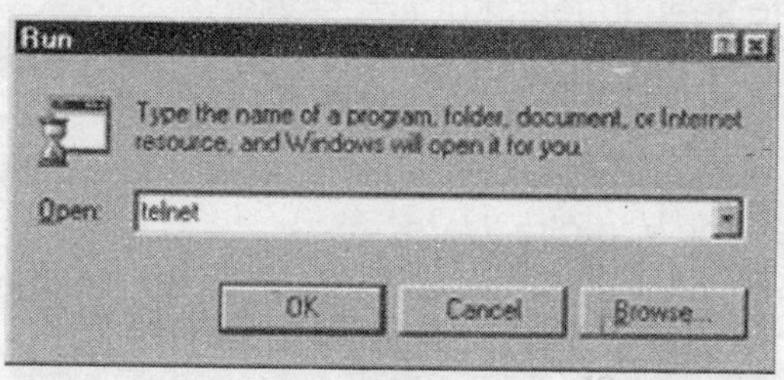

You should then see the blank Telnet window pop up. From the **Connect** menu, choose **Remote System....** You should then see a dialog box asking you for the **Host Name**, **Port** and **Term Type**.

In the **Host Name** box, enter the address of your web server, e.g. “www.mysite.com”. Leave the other settings as they are and press **Connect :**

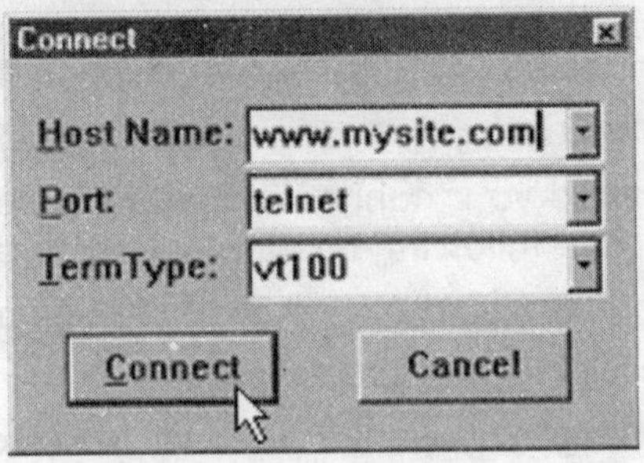

A short while after you click the **Connect** button, you should see some text appear in the main Telnet window, followed by the *login prompt*, similar to the following:

```
Kernel 2.2.14-12.8RS on an i686
login:
```

Now type in your user name, which should appear after the login: prompt, and press the Enter key. You should then see:

```
Password:
```

Enter your password, and press the Enter key again. If all goes well, you should see a short greeting message followed by the *shell prompt*, like this:

```
Last login: Mon Jun 18 10:56:42 from dialup123.isp.net
You have mail.
[username@mysite]$
```

9. What is face book application? How can you use it?

Ans. In short, it is your home on the web. Find out more you can do with Facebook and why you should use it.

- Facebook allows you to maintain a friends list and choose privacy settings to tailor who can see what on your profile
- Facebook allows you to upload photos and maintain photo albums that can be shared with your friends
- Facebook supports interactive online chat and the ability to comment on your friends “walls” in order to keep in touch or just say ‘hi’.
- Facebook supports groups and fan pages, allowing businesses to effectively use Facebook as a vehicle for social media marketing
- Facebook’s developer network delivers advanced functionality in the form of social apps like Flixster and Vampire Wars
- Facebook Connect allows websites to interact with Facebook and allows Facebook to be used as a universal login authentication service

  You can use face book by surfing the URL : http://www.facebook.com/

❑❑❑

B

APPENDIX

Glossary

A

Anonymous Proxy: This type of proxy server identifies itself as a proxy server

Apache: It is generally recognized as the world's most popular Web server (HTTP server).

B

BCC: Short for Blind Carbon Copy

Bridge: It is a device that allows you to segment a large network into two smaller

Broadband: It is a high-speed Internet connection that makes surfing the Web more enjoyable.

Browser: It is a program that our computer runs to communicate with web servers on the Internet

C

CC: Short for Carbon Copy contains the e-mail addresses for people other than the primary recipients.

CGI: The common gateway interface, the interface for invoking programs from web servers.

Chat Room: It is a virtual place on the Internet where people from all walks of life from around the world can get together in one place and textually chat with one another.

Chat Servers: These servers enable a large number of users to exchange information

Client-Side Scripting: The scripts on the page are processed by the individual web browser that requests the page.

Collaborative Computing: It is a term describing a variety of activities where people interact with one another using desktops, laptops, palmtops, and sophisticated digital cellular phones.

Computer Network: It is defined to be an interconnected collection of autonomous computers.

D

Distorting Proxy: This type of proxy server identifies itself as a proxy server, but make an incorrect original IP address available through the http headers.

E

E-mail Address: E-mail address is required to send and receive e-mail.

E-mail Attachment: Attachment contains the names of files that you may be sending

Email Bombing: It is characterized by abusers repeatedly sending an email message to a particular address at a specific victim site.

Email spamming: Is a variant of bombing; it refers to sending email to hundreds or thousands of users

E-mail: It is electronic transmission of messages over Internet.

F

File Transfer Protocol: It makes it possible to move one or more files securely between computers

Firewall: It is software/Hardware which enforces an access control policy between an internal network in an organization and the rest of the Internet.

G

GroupWare Server: It is software designed to enable users to collaborate

H

High Anonymity Proxy: This type of proxy server does not identify itself as a proxy server and does not make available the original IP address.

HTML: Hypertext markup language, the format fro web pages, provide both formatting and hyperlinking.

HTTP: Hyper Text protocol is not a protocol dedicated for e-mail communications, but it can be used for accessing our mailbox.

I

IIS: Internet Information Server is a set of Internet-based services for servers created by Microsoft for use with Microsoft applications.

IMAP: Internet Message Access Protocol is a standard protocol for accessing e-mail from our local server.

Internet: It is a network of networks, linking computers to computers sharing the TCP/IP protocols.

IPv4: Internet Protocol version 4 (IPv4) is the fourth revision in the development of the Internet Protocol (IP)

IPv6: The Internet Engineering Task Force (IETF) has designated IPv6 as the successor to version 4 for general use on the Internet.

ISDN: Integrated Services Digital Network is a high-speed dial-up

ISP: Internet Service Providers are an organization or business offering public access to the Internet.

J

JavaScript: It is a simple scripting language invented specifically for use in web browsers to make websites more dynamic.

L

Local Area Network (LAN): It is a network that is confined to a relatively small area.

M

Mail Servers: These serves move and store mail over corporate networks via LANs and WANs and across the Internet.

Mailer: Mailer is a company that sends direct mail.

message travels from sender to receiver.

Metropolitan Area Network: MAN is a computer network that is meant to span a whole metropolitan area.

Microsoft Front Page Express: FP Express is a cut down version of Microsoft's Web authoring software which provides a user-friendly interface to create web sites.

MIME: It is a standard for describing different types of information.

Modem: It is a Modulator/Demodulator; it connects a terminal/computer (DTE) to the Voice Channel (dial-up line).

N

Network Topology: It refers to the way in which a network is laid out physically

Newsgroup: It is an Internet community which is dedicated to the discussion of a particular topic of interest.

P

Packet Filtering: It is a technique whereby routers have ACLs (Access Control Lists) turned on. By default, a router will pass all traffic sent it, and will do so without any sort of restrictions.

Password: Password is a secret word or string of characters that is used for authentication

Plug-ins: These are software modules that add to the functionality of an application.

POP: Post Office Protocol provides a simple, standardized way for users to access mailboxes and download messages

Promotional mailers: These mailers are used for advertising and promotional purposes.

Protocol: A protocol is a set of rules that govern data communications.

Proxy Server: It is a server that acts between user's networks and the Internet for seeking resources from other servers".

Proxy Server: It is an application that mediates traffic between a protected network and the Internet.

Public-key encryption: This asymmetric encryption involves the use of two keys, one that can be used to encrypt messages (the public key) and one that can be used to either encrypt them or decrypt them (the private key).

PWS: an abbreviation for Personal Web Server, is Microsoft's version of a Web server program.

R

Repeaters: It is a device which regenerates incoming electrical, wireless or optical signals.

Ring Topology: In it topology devices are connected in a closed loop so that each device is connected to two others, one on either side.

Router: This device translates information from one network to another; it is similar to a super intelligent bridge.

S

Search Engine: A program that searches documents for specified keywords and returns a list of the documents where the keywords were found.

Secret-key encryption: This is also known as single-key or symmetric encryption, involves the use of a single key that is shared by both the sender and the receiver of the message.

Server: Server is computer, or a software package, that provides a specific kind of service to client software running on other computers.

Server-Side Scripts: These scripts are small programs that reside on the web server to interact with the web environment.

SGML: Standard Generalized Markup Language is an internationally agreed standard for information representation.

SMTP: Simple Mail Transfer Protocol is used by the Mail Transfer Agent (MTA) to deliver our e-mail to the recipient's mail server.

SSL: Secure Socket Layer is a protocol designed and implemented by Netscape Communications. At the socket layer, to protect any higher level protocol builds on sockets, such as Telnet, FTP and HTTP.

Star Topology: In this topology the devices are all connected to a central hub.

Switch: It is a small hardware device that joins multiple computers together.

T

Telnet Server: This enables users to log on to a host computer and perform tasks as if they're working on the remote computer itself.

Transmission Medium: The transmission medium is the physical path by which a

Transparent Proxy: This type of proxy server identifies itself as a proxy server and also makes the original IP address available through the http headers.

U

URLs: Uniform Resource Locators, are the method by which documents or data are addressed in the World Wide Web.

W

Web Server: It is the program resides on server computer to serve the request of the client software.

Wide Area Networks (WANs) : MAN connect larger geographic areas, such as Florida, the United States, or the world.

WWW: World Wide Web is broad term, which is the collection of distributed documents.

X

XML: It stands for Extensible Markup Language and it is a meta language to let users design their own markup language.

❑❑❑

EXAMINATION PAPERS (UNSOLVED)

Fourth Semester Examination
INTERNET FUNDAMENTALS (CSE-208-E)
(Years 2005-2006)

Note : Attempt any five questions. All questions carry equal marks.

Q. 1. (a) What is e-mail ? Why is it is popular ?

(b) Discuss different components of most common e-mail address format.

(c) Can you invoke a spell checker from inside your mailer? Describe how you go about doing this.

(d) Explain the difference between attaching a file and inserting a file in relation to an e-mail message.

Q. 2. What is Internet ? Provide an intuitive idea of how the internet functions.

Q. 3. (a) What are Plug-Ins? Justify their significance. Describe the process to download the plug-ins.

(b) What is meant by Metasearch Engine? How does a metasearch engine work?

Q. 4. (a) What factors determine how quickly a file transfer over the internet?

(b) Create a file sample-html having the basic tags that any HTML document should contain.

Q. 5. (a) What is XML? Write the reasons behind the emergence of XML. Also discuss application areas of XML.

(b) Using JAVA Script design a webpage that converts feet into inches and vice versa.

Q. 6. Write short notes on the following web servers.

(a) PWS (b) Microsoft Personal Web Server

Q. 7. (a) What are threats to internet security? Discuss.

(b) Describe Proxy Server. Discuss its role in internet security.

(c) What do you understand by encryption of Message? Why this concept came into existence ? Also write basic concept behind encryption schemes.

Q. 8. (a) What are Chat Rooms? How does chat work?

(b) What are e-mail mailing lists? How do you create them? Explain with the help of example.

Fourth Semester Examination
INTERNET FUNDAMENTALS (CSE-208-E)
(Years 2006-2007)

Note : Attempt any five questions. All questions carry equal marks.

Q. 1. (a) What is e-mail? Explain its internal working.

(b) Explain the following terms related to e-mail :

SMTP, MIME, PPP, POP3

Q. 2. (a) What is Internet ? How does it work? What are its advantages and disadvantages ?

(b) Differentiate between :

(i) Web page and home page

(ii) IPv4 and IPv6

Q. 3. (a) What is FTP? How is works?

(b) Explain the working of a search engine.

Q. 4. (a) Write a detailed note on XML.

(b) Write a note on server side programming in Javascript.

Q. 5. What are web servers? What is their utility? Discuss about IIS.

Q. 6. (a) Differentiate between the following:

(i) Authorization and Authentication

(ii) Privacy and secrecy

(b) Explain the following terms:

(i) Preventive maintenance

(ii) Troubleshooting

Q. 7. (a) What is an HTML document? Draw the structure of such document.

(b) What are different ways of connecting to Internet ?

Q. 8. Write the short notes on :

(a) Firewalls

(b) Digital signatures

(c) DNS

(d) ISP

❑❑❑

PREVIOUS YEARS SOLVED EXAMINATION PAPERS

FOURTH SEMESTER EXAMINATION, May 2007
INTERNET FUNDAMENTALS (CSE-208-E)

Note : Attempt any five questions. All questions carry equal marks

Q. 1. (a) Write down the advantages of the following in e-mail:

(i) address book

(ii) signature

(iii) attachment

Answer 1(a)

(i) E-mail Address Books : Your e-mail "address book" is the component of e-mail system which enables addition of *e-mail addresses* to your contact list quickly and without error. Your address book may store all of your frequently used e-mail addresses, enabling you to select from a list instead of having to re-type the addresses each time you send an e-mail. This is the fastest way to address a message, and more importantly ensures that you will never make a typing mistake and send your e-mail to the wrong address.

Address Book Advantages :

(a) E-mail programs have an **address book feature** where you can record the e-mail addresses and other information of people you contact frequently.

(b) By using the address book you won't have to retype an e-mail address each time you want to send a message. This **saves time** and **avoids returned mail** due to typing mistakes.

(c) With some programs, you can store **multiple e-mail addresses** for each person and additional information, such as a company name, telephone number, and street address.

(d) With most programs, you can create a set of e-mail addresses called a **group**. When you address a message with a group, the message is automatically **addressed to all the people** in the group. This is sometimes referred to as a **mailing list.**

(e) As your list grows, use the **Search** or **Find** function in the address book to locate an e-mail address based on a person's first or last name.

(ii) E-mail Signature : An e-mail signature is a small bit of information that one attaches to the end of an e-mail message. This information is typically your name and contact details; however, you can choose to put whatever you like. For instance, if you have a habit of signing off each e-mail with *God Bless, Regards, YOUR-NAME*, you can put this as an e-mail signature and it will be added at the end of all messages you compose.

E-mail signatures can be plain text or something fancier with images and colors. Any e-mail program worth its salt allows you to create these and the popular ones even let you have different signatures for each e-mail account you have configured in the software. For example, Outlook Express, which is a free e-mail program from Microsoft, lets you quickly create and attach e-mail signatures to outgoing messages.

The Importance and Advantages of E-mail Signatures : The main advantages of an e-mail signature are ease and time-saving. Even if you have a simple thank you followed by your name, imagine the convenience of having it automatically added to your e-mail message. However, e-mail signatures become much more useful once you put in your contact details including your website address, phone numbers and alternate e-mail addresses - the recipient will immediately know how to contact you!

Another way in which you can exploit the power of e-mail signatures is to mention ongoing promotions. So if you have a 20% off sale in progress, be sure to include it in the signature.

(iii) E-mail Attachments : It is possible to attach files to your emails. Click on the toolbar or icon (normally a paperclip), find the file you want to attach, and then click to attach it - one file successfully attached. You can repeat this operation to attach many files to a single e-mail.

Checking and removing attachments : You may sometimes want to check that the file you attached is the right one.

To check :

1. Right-click an *Attachment* file.
2. Choose *Quick View* from the pop-up menu.
3. A special viewer window will appear and you can check the document.
4. When you are happy close the *Quick View*.
5. If it is the wrong file, right clicking also provides the *Remove* option.

Advantages of E-Mail Attachment :

1. We can send text, video, audio, and multimedia files in attachments.
2. E-mail user can forward a complete e-mail message as attachments.
3. Attachments can be used in internet business to send the document related to business.
4. Attachment can help us in online advertising.
5. We can save lot of money, paper and time by using the e-mail attachments.

Q. 1. (b) Write an explanatory note on e-mail management.

Answer 1(b)

E-mail is now the de facto standard for business communication across organizations at this time. Recent research shows the extensive use of e-mails in business and its projected growth in the future presents major information management challenges and risks for many organizations. Just as any type of business information and record, e-mail must be included as part of, and adhered to the organizational standards addressing information and records:

- Capture
- Classification
- Storage

- Preservation
- Management
- Destruction

The vast quantities of e-mails held in inboxes, sent folders, and deleted item folders put the organization at risk and adversely impact the performance of e-mail servers in the organizations. E-mail servers were never designed to act as repositories for such great quantities of e-mails and move control of this information away from the organization. Without the management of e-mails, it is difficult for organizations to meet their legal preservation requirements in the event of litigation and government investigations, increasing the effort and cost in responding to e-discovery and disclosure.

E-mail management systems centrally capture e-mails created and received by employees. Using a classification scheme to manage this content, retention periods and access controls can be applied to manage e-mails. Metadata associated with the e-mails can be captured to allow this information to be managed and retrieved. E-mail active archiving is one of the most common applications for e-mail management. At their most basic, these solutions either copy or remove messages from the messaging application and some it at another location. Some approaches copy all messages coming into and out of the messaging application in real time, while others will physically remove the messages from the message store. In some of these solutions, the messages are not archived, but attachments are and are replaced with either a stub or an outright link within the message.

These e-mail management systems, while attractive and effective are merely part of an overall solution. Organizations need to have records management programs, consisting of organization-wide policies and procedures, staff and activities, in which these computer applications can be successfully utilized.

Q. 2. (a) What in Internet ? How is it related to Arpanet ? Write brief history of internet.

Answer 2(a)

Internet : ***The Internet*** *refers to millions of computers connected to a gigantic network and communicating via* ***TCP/IP protocols.***

A protocol is a pre-defined way for a computer to communicate with another computer, for instance when requesting a service, or when forwarding some information to another machine. Each computer at any given time has a unique address on the Internet. This is its IP address.

Arpanet and Internet History : Until the 1960s, very little communication was taking place between computers, and was carried out by the technology of telephone networks, i.e. circuit switching. In 1962 Paul Baran and Donald Davies independently proposed an idea of a robust, efficient, store-and-forward data network based on packets, i.e. units of data carried independently from one another. The technology was called **packet switching**, and has been implemented as ARPANET (a network of research sites in US, a project funded by the US Department of Defense) in late 1960s. In early 1970s ARPANET spanned the continental US, and by 1973 it had connections to Europe.

ARPANET had several protocols for communications of computers one to another or with the network. In 1974 a new, more robust suite of protocols has been developed and implemented on the ARPANET based on **Transmission Control Protocol** (TCP) and **Internet Protocol**

(IP). The standard includes a large collection of protocols, some of which we mention later. These protocols have been modified several times since then, but essential ideas of the original protocols are still preserved. In 1983 the US Department of Defense has mandated that all of their computer systems use TCP/IP protocols, which boosted use of these protocols in the US and throughout the world. Another boost to TCP/IP was given by including it, also in 1983, into the communication kernel of the University of California's UNIX implementation.

In 1986 the National Science Foundation (NSF) built a backbone network (i.e. a network of fast powerful computers which can quickly forward information to each other and other computers) called NSFNET. Eventually it grew to provide connectivity between various networks, forming the foundation of what currently is known as Internet.

In 1993 the NSF started to reduce its role in governing the Internet. Currently the administration of the Internet is divided between several international organizations, such as ISOC (Internet Society) and IAB (Internet Activities Board), which supervise various technical, administrative, and other aspects of the Internet.

Q. 2. (b) What is e-commerce? What are its types ?

Answer 2(b)

Electronic Commerce Defintion : E-commerce is the use of Internet and the web to transact business but when we focus on digitally enabled commercial transactions between and among organizations and individuals involving information systems under the control of the firm it takes the form of e-business. Nowadays, 'e' is gaining momentum and most of the things if not everything is getting digitally enabled. Thus, it becomes very important to clearly draw the line between different types of commerce or business integrated with the 'e' factor.

There are mainly five types of e-commerce models:

1. **Business to Consumer (B2C) :** As the name suggests, it is the model involving businesses and consumers. This is the most common e-commerce segment. In this model, online businesses sell to individual consumers. When B2C started, it had a small share in the market but after 1995 its growth was exponential. The basic concept behind this type is that the online retailers and marketers can sell their products to the online consumer by using crystal clear data which is made available via various online marketing tools. E.g. An online pharmacy giving free medical consultation and selling medicines to patients is following B2C model.
2. **Business to Business (B2B)** - It is the largest form of e-commerce involving business of trillions of dollars. In this form, the buyers and sellers are both business entities and do not involve an individual consumer. It is like the manufacturer supplying goods to the retailer or wholesaler. E.g. Dell sells computers and other related accessories online but it is does not manufacture all those products. So, in order to sell those products, it first purchases them from different businesses i.e. the manufacturers of those products.
3. **Consumer to Consumer (C2C)** - It facilitates the online transaction of goods or services between two people. Though there is no visible intermediary involved but the parties cannot carry out the transactions without the platform which is provided by the online market maker such as eBay.
4. **Peer to Peer (P2P)** - Though it is an e-commerce model but it is more than that. It is a technology in itself which helps people to directly share computer files and computer resources without having to go through a central web server. To use this, both sides need

to install the required software so that they can communicate on the common platform. This type of e-commerce has quite low revenue generation as from the beginning it has been inclined to the free usage due to which it sometimes got entangled in cyber laws.

5. **M-Commerce** - It refers to the use of mobile devices for conducting the transactions. The mobile device holders can contact each other and can conduct the business. Even the web design and development companies optimize the websites to be viewed correctly on mobile devices.

There are other types of e-commerce business models too like Business to Employee (B2E), Government to Business (G2B) and Government to Citizen (G2C) but in essence they are similar to the above mentioned types. Moreover, it is not necessary that these models are dedicatedly followed in all the online business types. It may be the case that a business is using all the models or only one of them or some of them as per its needs.

Q. 3. (a) Draw and explain the conceptual organization of a browser.

Answer 3(a)

Web Browser : Surfing the web is made possible by **Web browsers**. Browsers are basically software programs that allow you to search for and view various kinds of information on the Web, such as web sites, video, audio, etc.

Here are just a few of Web browsers available to you for a free download:

1. Microsoft's Internet Explorer
2. Opera
3. Mozilla's Firefox
4. Mac Safari

The reference architecture we derived is shown in Fig. 1. It comprises eight major subsystems plus the dependencies between them :

(1) **The User Interface subsystem** is the layer between the user and the Browser Engine. It provides features such as toolbars, visual page-load progress, smart download handling, preferences, and printing. It may be integrated with the desigktop environment to provide browser session management or communication with other desktop applications.

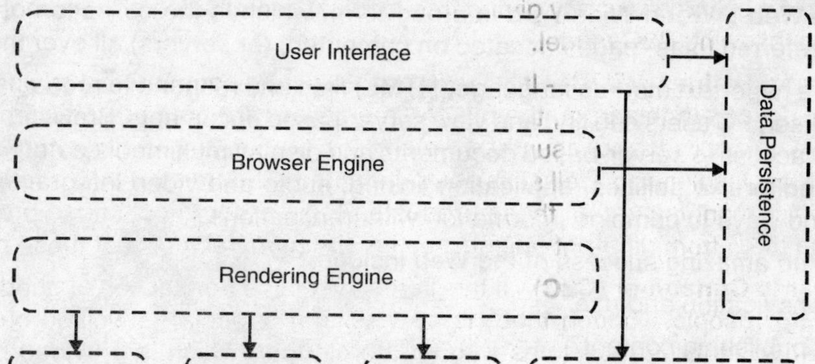

Fig. 1. Reference Architecture for Web Browsers.

(2) **The Browser Engine subsystem** is an embeddable component that provides a high-level interface to the Rendering Engine. It loads a given URI and supports primitive browsing actions such as forward, back and reload. It provides hooks for viewing various aspects of the browsing session such as current page load progress and JavaScript alerts. It also allows the querying and manipulation of Rendering Engine settings.

(3) The Rendering Engine subsystem produces a visual representation for a given URI. It is capable of displaying HTML and Extensible Markup Language (XML) documents, optially styled with CSS, as well as embedded content such as images. It calculates the exact page layout and may use "reflow" algorithms to incrementally adjust the position of elements on the page. This subsystem also includes the HTML parser.

(4) **The Networking subsystem** implements file transfer protocols such as HTTP and FTP. It translates between different character sets, and resolves MIME media types for files. It may implement a cache of recently retrieved resources.

(5) **The *JavaScript Interpreter* evaluates JavaScript** (also known as ECMA-Script) code, which may be embedded in web pages. JavaScript is an object-oriented scripting language developed by Netscape. Certain Java-Script functionality, such as the opening of pop-up windows, may be disabled by the Browser Engine or Rendering Engine for security purposes.

(6) **The XML Parser subsystem parses** XML documents into a Document Object Model (DOM) tree. This is one of the most reusable subsystems in the architecture. In fact, almost all browser implementations leverage an existing XML Parser rather than creating their own from scratch.

(7) **The Display Backend subsystem provides** drawing and windowing primitives, a set of user interface widgets, and a set of fonts. It may be tied closely with the operating system.

(8) **The Data Persistence subsystem stores** various data assocaited with the browsing session on disk. This may be high-level data such as bookmarks or toolbar settings, or it may be low-level data such as cookies, security certificates, or cache.

Q. 3. (b) What is WWW? How are the addresses defined on Internet?

Answer 3(b)

World Wide Web : World Wide Web is broad term, which is the collection of distributed documents referred to as "pages" located on computers (or servers) all over the world.

Servers store hypertext markup language (HTML) files and respond to requests. Through the use of a browser, PC users can find and view server-based documents. Browsers like Netscape ensure easy access to server-based documents and display multimedia data. Future browser will include editor-like abilities, application linking, audio and video integration; data based front end, and ways to combine information with transactions.

Reason for the amazing success of the Web include:

- Ease of navigation and use.
- Ease of publishing content.
- New distribution models.
- Enabling a network-centric computing paradigm.
- Enabling new intra-business applications.

Internet Addressing Schemes : As internet is a network of networks, it basically also consists of network nodes like computers. All these devices are interconnected to communicate with each other. Each device in a network must have a unique address to be identified. This address makes this device unique on that network and any other computer or device can communicate with it by using this specific address.There are two major types of addressing schemes that are used in the internet environment:

1. IP addressing
2. DNS addressing

In the IP addressing scheme IP stands for internet protocols. IP is a number or value that is used to uniquely identify a computer on the internet. It is a 32-bit value and this number/value can be divided into four different sections. All these sections are separated by a period. As each section is of 8 bits thus, it can represent a value in the range of 0 to 255.

Each section is known as an octet. Each time a computer is connected to the internet, it is assigned a specific IP address dynamically by the ISP. The DNS stand for domain name system. Unlike the IP addressing scheme in this addressing scheme text is used instead of numbers. It is easier to remember than IP address.

Q. 4. (a) Discuss structure of an HTML document.

Answer 4(a)

HTML : HTML, which stands for Hypertext Markup Language, is a markup language used to create web pages. The web developer uses "HTML tags" to format different parts of the document. For example, you use HTML tags to specify headings, paragraphs, lists, tables, images and much more.

HTML is a subset of Standard Generalized Markup Language (SGML) and is specified by the World Wide Web Consortium (W3C).

Structure of an HTML program : Every HTML program has one simple structure. There are two major components to the structure of a website, the "head" and "body". The "head" provides the title and other information about your website , while the "body" provides the information and content. Each of these components are marked by "tags", which let the computer know where the information is written. There are two types of "tags", opening tags and closing tags. Opening tags have one triangular bracket (</>) on each of their two sides, while closing tags start and end with a triangular bracket and a slash (/)at the beginning. Lastly, all web sites begin with the tag <html> and close with the tag </html>.

Example:

```
<html>

<head>
<title>|
</title>
</head>

<body>
</body>

</html>
```

Q. 4. (b) What is XML? Where is it used ?

Answer 4(b)

XML : XML is a highly functional subset of SGML. The purpose of XML is to specify an SGML subset that works very well for delivering SGML information over the Web. When the mainstream Web browsers support XML, it is believed that it's going to be very easy to publish SGML information on the Web.

XML stands for Extensible Markup Language and it is a markup language much like HTML. It is a meta language to let users design their own markup language and was designed to carry data, not to display data. XML tags are not predefined and You must define your own tags.XML is designed to be self-descriptive.

Use of XML : XML is used in many aspects of web development, often to simplify data storage and sharing.

1. **XML Separates Data from HTML :** If you need to display dynamic data in your HTML document, it will take a lot of work to edit the HTML each time the data changes. With XML, data can be stored in separate XML files. This way you can concentrate on using HTML for layout and display, and be sure that changes in the underlying data will not require any changes to the HTML.
2. **XML Simplifies Data Sharing :** In the real world, computer systems and databases contain data in incompatible formats.XML data is stored in plain text format. This provides a software- and hardware-independent way of storing data. This makes it much easier to create data that different applications can share.
3. **XML Simplifies Data Transport :** With XML, data can easily be exchanged between incompatible systems. One of the most time-consuming challenges for developers is to exchange data between incompatible systems over the Internet. Exchanging data as XML greatly reduces this complexity, since the data can be read by different incompatible applications.
4. **XML Simplifies Platform Changes :** Upgrading to new systems (hardware or software platforms), is always very time consuming. Large amounts of data must be converted and incompatible data is often lost.XML data is stored in text format. This makes it easier to expand or upgrade to new operating systems, new applications, or new browsers, without losing data.
5. **XML Makes Your Data More Available :** Since XML is independent of hardware, software and application, XML can make your data more available and useful. Different applications can access your data, not only in HTML pages, but also from XML data sources. With XML, your data can be available to all kinds of "reading machines" (Handheld computers, voice machines, news feeds, etc), and make it more available for blind people, or people with other disabilities.
6. **XML is Used to Create New Internet Languages :** A lot of new Internet languages are created with XML. Here are some examples:

   (a) XHTML the latest version of HTML

   (b) WSDL for describing available web services

   (c) WAP and WML as markup languages for handheld devices

   (d) RSS languages for news feeds

   (e) RDF and OWL for describing resources and ontology

   (f) SMIL for describing multimedia for the web

Q. 5. (a) State significance of a web server. What is PWS ?

Answer 5(a)

Significance of Web Server : Web server is the program resides on server computer to server the request of the client software.

Web servers are a combination of software and hardware that store documents and other content. The server content is accessed using Web browsers. Web servers are available for a wide range of platforms including IBM PCs, Apple Macintosh computers, UNIX platforms and a large number of proprietary hosts.

Some the example of the web servers are as follows:

1. Tomcat
2. Weblogic
3. Apache
4. IIS

PWS : PWS, an abbreviation for Personal Web Server, is Microsoft's version of a Web server program for individual PC users who want to share Web pages and other files from their hard drive. PWS is a scaled-down version of Microsoft's more robust Web server, Internet Information Server IIS. PWS can be used with a full-time Internet connection to serve Web pages for a Website with limited traffic. It can also be used for testing a Website offline or from a "staging" site before putting it on a main Website that is exposed to larger traffic.

Here's what the Personal Web Server does for us :

- Allows webs to be published as a whole, with a single command.
- Publish from local to remote, remote to local, or remote to remote websites.
- Provides the functionality of CGI, IDC, ASP, ISAPI programs, and the FrontPage WebBots on your local computer, so that you can test your webs locally in your browser before making them public.
- Provides "smart" web management.
- Set permissions and properties of webs and folders.
- Sets up virtual directories on the server.
- You can actually browse your webs from another computer on the Internet, when you are logged on. This can be handy for demonstrating them to clients, friends, etc.

Q. 5. (b) Write a detailed note on Apache server.

Answer 5(b)

Apache : The Apache HTTP Server Project is a collaborative software development effort aimed at creating a robust, commercial-grade, and freely-available source code implementation of an HTTP (Web) server. The project is jointly managed by a group of volunteers located around the world, using tne Internet and the Web to communicate, plan, and develop the server and its related documentation. This project is part of the Apache Software Foundation. In addition, hundreds of users have contributed ideas, code, and documentation to the project. This file is intended to briefly describe the history of the Apache HTTP Server and recognize the many contributors.

Apache is generally recognized as the world's most popular Web server (HTTP server). Apache has attracted so much interest because it is full-featured, reliable, and free. Originally developed for UNIX™ operating systems, Apache has been updated to run on Windows, OS/2, and other platforms.

Features of Apache Web Server : The features of this server are as follows:

1. Is a powerful, flexible, HTTP/1.1 compliant web server
2. Implements the latest protocols, including HTTP/1.1 (RFC2616)
3. Is highly configurable and extensible with third-party modules
4. Can be customized by writing 'modules' using the Apache module API
5. Provides full source code and comes with an unrestrictive license
6. Runs on Windows 2003/XP/2000/NT/9x, Netware 5.x and above, OS/2, and most versions of Unix, as well as several other operating systems
7. Is actively being developed
8. Encourages user feedback through new ideas, bug reports and patches

Q. 6. What are various encryption standards?

Answer 6

Encryption : Encryption is the process used to hide our data, or the contents of a message, from prying eyes throughout the internet. During transmission (such as through a secure socket layer), the data is disguised using codes so that no one along the chain of networks that the data passes though to get to its source can understand the information being sent. When the data arrives at its destination, it is decrypted to reveal the information being transmitted.

The process of disguising the data is called encryption and the process of revealing the data from its encrypted form is called decryption. Both of these are common techniques used in cryptography - the scientific discipline behind secure connections. The processes are done using mathematical logic, or algorithms.

The main purpose of encryption algorithms is to provide the following:

Authentication : Proving one's identity before granting access.

Privacy and confidentiality : Ensuring that outsiders cannot read data intended for specific parties.

Integrity : Ensuring that the message has not be modified in any way before it arrives to the intended recipient.

Non-repudiation : Ensuring that a message is truly originated from the sender.

Data Encryption Standard (DES) : Data Encryption Standard (DES) is a block cipher with 64-bit block size that uses 56-bit keys. DES was invented over 20 years ago by IBM in response to a public request from the National Bureau of Standards. Due to recent advances in computer technology, some experts no longer consider DES secure against all attacks; since then Triple-DES (3DES) has emerged as a stronger method. Using standard DES encryption, Triple-DES encrypts data three times and uses a different key for at least one of the three passes giving it a cumulative key size of 112-168 bits.

1. **IDEA :** IDEA stands for International Data Encryption Algorithm. IDEA is a symmetric encryption algorithm that was developed by Dr. X. Lai and Prof. J. Massey to replace the DES standard. Unlike DES though it uses a 128 bit key. This key length makes it impossible to break by simply trying every key.
2. **RC2 :** RC2 is a variable-key-length cipher. It was invented by Ron Rivest for RSA Data Security, Inc. Its details have not been published.
3. **RC4 :** RC4 was developed by Ron Rivest in 1987. It is a variable-key-size stream cipher. It is a cipher with a key size of up to 2048 bits (256 bytes).
4. **RC6 :** RC6 is a symmetric key block cipher derived from RC5. It was designed by Ron Rivest, Matt Robshaw, Ray Sidney, and Yiqun Lisa Yin to meet the requirements of the Advanced Encryption Standard (AES) competition. RC6 encryption algorithm was selected among the other finalists to become the new federal Advanced Encryption Standard (AES).
5. **Serpent :** Serpent is a block cipher developed by Ross Anderson, Eli Biham and Lars Knudsen. Serpent can work with different combinations of key lengths. Serpent was also selected among other five finalists to become the new federal Advanced Encryption Standard (AES).
6. **Triple DES :** Triple DES is a variation of Data Encryption Standard (DES). It uses a 64-bit key consisting of 56 effective key bits and 8 parity bits. The size of the block for Triple-DES is 8 bytes. Triple-DES encrypts the data in 8-byte chunks. The idea behind Triple DES is to improve the security of DES by applying DES encryption three times using three different keys.
7. **Twofish :** Twofish is a symmetric block cipher. Twofish has a block size of 128 bits and accepts keys of any length up to 256 bits.

   Twofish encryption algorithm was designed by Bruce Schneier, John Kelsey, Chris Hall, Niels Ferguson, David Wagner and Doug Whiting. The National Institute of Standards and Technology (NIST) investigated Twofish as one of the candidates for the replacement of the DES encryption algorithm.

Q. 7. Write short notes on the following : (a) Java Script

Answer 7(a)

JavaScript : When the World Wide Web was first created in the early 1990s all web pages were static. When you viewed a web page you saw exactly what the page was set up to show you and there was no way for you to interact with the page.

Being able to interact with a web page - have it do something in response to your actions - required the addition of some form of programming language to "instruct" the page how it should respond to your actions. In order to have it respond immediately without having to reload the web page this language needed to be able to run on the same computer as the browser displaying the page.

At the time there were two browsers that were reasonably popular - Netscape Navigator and Internet Explorer. Netscape was the first to bring out a programming language that would allow web pages to become interactive - they called it **LiveScript** and it was integrated into the browser (meaning that the browser would interpret the commands directly without requiring the code to be compiled and without requiring a plugin to be able to run it). This meant that anyone using the latest Netscape browser would be able to interact with pages that made use of this language.

Another programming language called Java (which required a separate plugin in order to run) became very well known and so Netscape decided to try to cash in on this by renaming the language built into their browser to **JavaScript**. Note that while some Java and JavaScript code may appear similar, they are in fact two entirely different languages that serve completely different purposes.

JavaScript is a simple scripting language invented specifically for use in web browsers to make websites more dynamic.

JavaScript is an easy-to-use programming language that can be embedded in the header of your web pages. It can enhance the dynamics and interactive features of your page by allowing you to perform calculations, check forms, write interactive games, add special effects, customize graphics selections, and create security passwords and more.

JavaScript is most commonly used as a client side scripting language. This means that JavaScript code is written into an HTML page. When a user requests an HTML page with JavaScript in it, the script is sent to the browser and it's up to the browser to do something with it.

JavaScript is a feature rich and useful browser based script that if used properly can achieve some great effects and improve the experience for the end user. There are drawbacks to its use in that it tends to bloat web pages. The key thing to remember is to get the best of both worlds by using JavaScript code in external script files. That way the code is separated from your content so you get all the benefit of JavaScript functionality but without the adverse effects of the associated code bloating.

Advantages of JavaScript :

- **Speed :** Being client-side, JavaScript is very fast because any code functions can be run immediately instead of having to contact the server and wait for an answer.
- **Simplicity :** JavaScript is relatively simple to learn and implement.
- **Versatility :** JavaScript plays nicely with other languages and can be used in a huge variety of applications. Unlike PHP or SSI scripts, JavaScript can be inserted into any web page regardless of the file extension. JavaScript can also be used inside scripts written in other languages such as Perl and PHP.
- **Server Load :** Being client-side reduces the demand on the website server.

Disadvantages of JavaScript :

- **Security :** Because the code executes on the users' computer, in some cases it can be exploited for malicious purposes. This is one reason some people choose to disable JavaScript.
- **Reliance on End User :** JavaScript is sometimes interpreted differently by different browsers. Whereas server-side scripts will always produce the same output, client-side scripts can be a little unpredictable. Don't be overly concerned by this though - as long as you test your script in all the major browsers you should be safe

Q. 7. (b) Write short note on : Firewalls

Answer 7(b)

A *Firewall* is a computer program that monitors the flow of information from the Internet to your computer. There are two different types of firewall available for you to use - *Hardware Firewalls* and *Software Firewalls*.

Hardware Firewall : A *Hardware Firewall* is a physical piece of equipment that sits between the Internet and your computer. An example of a hardware firewall is a broadband router, a common form of Internet connection. The benefit of using a hardware firewall, is that it has the ability to protect multiple computer systems that are connected to it at the same time. This makes it an effective firewall for use in businesses that have multiple computers connected to the Internet, as well as in homes that have more than one computer system.

Software Firewall : *Software Firewalls* work in the same way as a hardware firewall, by monitoring and blocking information that comes to your computer via the Internet, however software firewalls must be installed as a program on your computer. These software firewalls can either be installed from a computer disk that you have purchased, or downloaded over the Internet. Software firewalls are the most common type of firewall. Programs such as Norton 360, Norton Internet Security, ESET Smart Security, and Kaspersky Internet Security all have a firewall bundled within them.

What a Firewall does : Regardless of what type of firewall you choose to use, having one on your computer is a really good idea. A firewall helps to prevent computer hackers from accessing your computer through the Internet, and stealing sensitive information or infecting your computer system with some form of computer virus. A hacker can gain access to your system by "back doors" or open ports that connect your computer to the Internet. With a firewall in place these ports are blocked from inbound traffic, effectively closing the door in the hackers face.

Of course, damage can also be done to your computer from outbound traffic as well. Say for instance a hacker was able to slip by you with a Trojan, virus, spyware or phishing scheme that went undetected. The next time you booted up your computer and connected to the Internet, information on your computer could be sent out to the hacker. Many firewalls also block outbound traffic that will help prevent this from happening.

There are basically two ways that Firewalls work. Generally, data that comes in is analyzed by the firewall to determine the IP address it is coming from and the content that it contains. The firewall system then checks to see if this information is compliant with rules that you are able to configure. It can also analyze information at the application level. The firewall program will determine whether or not the application should be able to send or receive data through the port you are connected by.

Firewall programs are extremely important to the protection of not only your computer system, but you personal information as well. Many anti-virus programs come with firewall protection. Firewall protection should be considered an essential part of any computer's security strategy.

FOURTH SEMESTER EXAMINATION, Dec. 2008

INTERNET FUNDAMENTALS (CSE-208-E)

Note : Attempt any five questions. All questions carry equal marks

Q. 1.(a) Explain the terms E-mail and Bouncing feature of E-mail. Give major reasons for a bounce.

Answer 1(a)

E-mail : You can send e-mail from your computer at any time of the day to any address around the world and your electronic letter will arrive at its destination seconds after you send it, even if the receiver lives on the other side of the world.

"*E-mail is as simple and straight forward as letter writing. E-mail is an electronic version of sending a letter*".

"*E-mail, e-mail or electronic mail is the transmission of messages (e-mails or e-mail messages) over electronic networks like the internet*".

Bouncing feature of E-mail : Mail delivery program attempts to connect to the recipient mail server. If the mail server is too busy, or doesn't like the way your system is talking to it, or has a list saying it shouldn't listen to your system at all, it will reject your connection. This is bouncing feature of e-mail.

There are two kinds of e-mail bounces:

A hard bounce is an e-mail message that has been returned to you because the recipient's address is invalid. A hard bounce might occur because the domain name doesn't exist, the recipient is unknown, or there's some type of network problem on the recipient's end.

A soft bounce is an e-mail message that gets as far as the recipient's mail server, but is bounced back before it reaches the recipient. One of the most common causes for a soft bounce is a full mailbox. This will happen A LOT with your subscribers who use free e-mail services like Yahoo and HotMail, because they allow for very little e-mail storage.

Major Reasons of Bouncing

1. Mailbox Not Found, invalid mailbox, User unknown, not our customer: these are all saying pretty much the same thing. In the "someone@example.com" bounce examples above, the mail server "example.com" doesn't have an account for anyone with the e-mail name "someone". A couple of common reasons:
   (a) You typed the e-mail address wrong. The single most common reason this error happens is simply that you made a typographical error in the e-mail name. Check the entire e-mail address for an error.
   (b) It's an old e-mail address that's no longer in use. Perhaps the person you're attempting to e-mail has changed their e-mail address and you're using an old one which is no longer valid. Make sure what you're using is up to date.

2. **Mailbox unavailable:** 9 times out of 10, this is the same as "mailbox not found". That other 10% of the time it could mean that there's a problem with the recipients e-mail account. What kind of problem is hard to say. Check to make sure that you have the e-mail address correct, wait a while and try again, and if it still bounces try contacting the recipient some other way.
3. **Mailbox full, or Quote Exceeded:** Sometimes this will show up as a part of a "Mailbox unavailable" message. It's fairly clear, though: your recipient has too much e-mail and their server isn't accepting any more. This is most common with web-based e-mail services like Hotmail or Yahoo, which have limits on how much mail you can accumulate.
4. **Host unknown, Domain Lookup Failed:** This means that the mail server you're attempting to use, the "example.com" part, in the examples above, doesn't exist. A common reason is again, a typo on your part. Make sure you typed it in correctly. Another reason are ISPs that change their name.
5. **Unable to Relay:** This is a terribly obscure error message, but also becoming more and more common as ISPs try to crack down on spam. Mail is sent by relaying e-mail from one server to the next. There could be many servers involved, but typically it's the mail server at your ISP relaying your e-mail to the mail server at your recipients ISP.
6. **Temporary Errors:** errors like "no adequate servers", "Connection Timed Out", "Resources temporarily unavailable.", "Out of memory" all typically indicate a problem with a mail server that you probably don't have any control over. They are, in general, temporary, and should resolve themselves over time. *Look carefully at the bounce message*; the e-mail server involved may continue to automatically try to deliver your e-mail without any action required on your part.
7. **Blacklist Filters:** If you see messages that indicate your e-mail was "blocked", or "listed in", and references to sites that have things like "spamcop", "dynablock", "blackhole", "spamhaus" and similar in their names, then your e-mail was probably intentionally blocked because the receiving system thinks your ISP's mail server is a source of spam.
8. **Content Filters:** Much like blacklists, content filters are an approach many ISPs now implement to stem the tide of spam for their clients. Most will simply discard e-mail that looks like spam.

Q. 1. (b) Explain SMTP, POP, Central Mail Spool, and IMAP in context of E-mail.

Answer 1(b)

SMTP : *SMTP is the acronym for "Simple Mail Transfer Protocol", which is the standard technology used by people around the globe when sending e-mail.*

Working of SMTP : SMTP is a simple ASCII protocol. After establishing the TCP connection to port 25, the sending machine, operating as the client, waits for the receiving machine, operating as the server, to talk first. The server starts by sending a line of text giving its identity and telling whether or not it is prepared to receive mail. If it is not, the client releases the connection and tries again later.

If the server is willing to accept e-mail, the client announces whom the e-mail is coming from and whom it is going to. If such a recipient exists at the destination, the server gives the client the go-ahead message. Then the client sends the message and the server acknowledges it.

IMAP (Internet Message Access Protocol)

*IMAP stands for **I**nternet **M**essage **A**ccess **P**rotocol. It is a method of accessing electronic mail or bulletin board messages that are kept on a (possibly shared) mail server.*

In other words, it permits a "client" e-mail program to access remote message stores as if they were local. For example, e-mail stored on an IMAP server can be manipulated from a desktop computer at home, a workstation at the office, and a notebook computer while traveling, **without** the need to transfer messages or files back and forth between these computers.

IMAP's ability to access messages (both new and saved) from more than one computer has become extremely important as reliance on electronic messaging and use of multiple computers increase, but this functionality cannot be taken for granted: the widely used Post Office Protocol (POP) works best when one has only a single computer, since it was designed to support "offline" message access, wherein messages are downloaded and then deleted from the mail server. This mode of access is not compatible with access from multiple computers since it tends to sprinkle messages across all of the computers used for mail access. Thus, unless all of those machines share a common file system, the offline mode of access that POP was designed to support effectively ties the user to one computer for message storage and manipulation.

Key goals for IMAP include:

- Be fully compatible with Internet messaging standards, e.g. MIME.
- Allow message access and management from more than one computer.
- Allow access without reliance on less efficient file access protocols.
- Provide support for "online", "offline", and "disconnected" access modes
- Support for concurrent access to shared mailboxes
- Client software needs no knowledge about the server's file store format.

POP (Post Office Protocol) :

Protocol provides a simple, standardized way for users to access mailboxes and download messages to their computers. When using the POP protocol all your e-mail messages will be downloaded from the mail server to your local computer.

The advantage is that once your messages are downloaded you can cut the internet connection and read your e-mail at your leisure without incurring further communication costs. On the other hand you might have transferred a lot of message (including spam or viruses) in which you are not at all interested at this point.

What the Post Office Protocol Allows You to Do

Like it seems everything on the internet, mail retrieval is a client-server application. The Post Office Protocol defines how your e-mail client should talk to the POP server. The POP is a very simple protocol. This makes it easy to implement, has earned the Post Office Protocol widespread adoption and makes it very robust, but it also means the Post Office Protocol provides only basic functionality.

Things that can be done via the POP include:

- Retrieve mail from an ISP and delete it on the server.
- Retrieve mail from an ISP but not delete it on the server.

- Ask whether new mail has arrived but not retrieve it.
- Peek at a few lines of a message to see whether it is worth retrieving.

Of these, the second probably sounds the most dangerous. Deleting something is always frightening. Remember, though, that you retrieve your mail before you delete it and thus have a copy.

Central Mail Spool : When someone sends mail to your @stanford e-mail address, the message actually gets sent to a buffer facility on the concern e-mail server and this facility is called central mail spool. It waits there, along with any other messages that come in, until you check your e-mail.

When you check your e-mail, the message is sent from your mail spool to your e-mail program. If your e-mail program uses the POP protocol, the message is downloaded onto your desktop computer. If you use a UNIX-based program like Pine or Elm, the message is downloaded into the mail directory of your Stanford account. This clears your mail spool.

But, if your e-mail program has been set to leave a copy of your e-mail on the server or your e-mail program is configured for IMAP, your mail stays on the mail spool until you remove it. Eventually, the size of your mail spool may prevent your e-mail program from working correctly.

Q. 2. (a) Describe HTML document along with 10 different HTML tags.

Answer 2(a)

HTML : HTML, which stands for Hypertext Markup Language, is a markup language used to create web pages. The web developer uses "HTML tags" to format different parts of the document. For example, you use HTML tags to specify headings, paragraphs, lists, tables, images and much more.

HTML is a subset of Standard Generalized Markup Language (SGML) and is specified by the World Wide Web Consortium (W3C).

Structure of an HTML program : Every HTML program has one simple structure. There are two major components to the structure of a website, the "head" and "body". The "head" provides the title and other information about your website , while the "body" provides the information and content. Each of these components are marked by "tags", which let the computer know where the information is written. There are two types of "tags", opening tags and closing tags. Opening tags have one triangular bracket (</>) on each of their two sides, while closing tags start and end with a triangular bracket and a slash (/)at the beginning. Lastly, all web sites begin with the tag <html> and close with the tag </html>.

Example:

```
<html>

<head>
<title>|
</title>
</head>

<body>
</body>

</html>
```

HTML Tags : HTML works in a very simple, very logical, format. It reads like you do, top to bottom, left to right. **HTML** is written with **TEXT**. What you use to set certain sections apart as bigger text, smaller text, bold text, underlined text, is a series of tags.

Think of tags as commands. For example if you want a line of text to be bold. You will put a tag at the exact point you want the bold lettering to start **<bold>and another tag where you want the bold lettering to stop.</bold>** There you have an example of how tags make commands and tell the browser how to display the text. Just like telling a dog to sit. You give the command.

All tags that are opened must correspondingly be closed, just as if you are writing a quoted statement with those "inverted commas". A tag is closed this way **</tag>** therefore we open a new tag **<bold> and close the tag </bold>**

Here is the summary of some basic tags:

Tags	Description
<html></html>	Tags which create the HTML document.
<head></head>	Tags which denote the first part of the HTML document.
<title></title>	Tags which define the title of the web page.
<body></body>	Tags which define the main contents of the HTML document.
<h1></h1>	Tags that indicate the headings. There are 6 headings, from h1 to h6.
<b></b>	Tags that display **bold** text.
<i></i>	Tags that display *italic* text.
<u></u>	Tags that underline the text.
 	Tag that indicate line break. No closing tag is required here.

Q. 2. (b) Differentiate between Get and Post Method.

Answer 2(b)

GET and POST methods are both ways of transporting variables (and their values) from one page to another. However, they transport the variable names and values in different ways.

GET uses the URL to transfer variables. Web pages can pick up the URL address and detach the variables usable to the scripting language used in the page. A user can change the variable names and values of a URL with GET-type variables. This introduces security risks, as it is an open way to attack a website. All variables that could possibly be input by a user should always be checked. GET values must be under 255 characters, otherwise some older browsers will not transfer values properly. Today's browsers propagate values just fine with larger character limits, such as 4,000 - but caution should be taken.

POST transfers variables from one page to the next - broadly speaking - silently. No users can edit these variables without simulating a POST transfer to the page - which is a more difficult route for site attackers to make. These variables should still always be checked. POST values can be of infinite length.

Q. 3. What do you understand by Encryption schemes? Explain any three encryption scheme.

Answer 3.

Encryption schemes : Encryption is the process used to hide our data, or the contents of a message, from prying eyes throughout the internet. During transmission, the data is disguised using codes so that no one along the chain of networks that the data passes though to get to its source can understand the information being sent. When the data arrives at its destination, it is decrypted to reveal the information being transmitted.

Fig. 2. Secret Key Encryption

The process of disguising the data is called encryption and the process of revealing the data from its encrypted form is called decryption. Both of these are common techniques used in cryptography - the scientific discipline behind secure connections. The processes of doing encryption is called encryption scheme.

The Three main encryption schemes are as follows:

1. **Secret-key encryption :** Secret-key encryption, also known as single-key or symmetric encryption, involves the use of a single key that is shared by both the sender and the receiver of the message.

   After creating the message, the sender encrypts it with their key and passes it to the recipient who then decrypts it by using a copy of the same key used to encrypt it.

   A widely used method of secret-key encryption is the data encryption standard or DES. **Secret-key encryption does have some limitations,** particularly with regard to key distribution.

   (a) For privacy to be maintained, every transmitter of messages would need to provide a different key to everyone they intended to communicate with, otherwise every potential recipient would be able to read all messages whether it was intended for them or not.

   (b) While this is manageable where a small number of parties are involved (for example, sending a private e-mail to a friend) it is not practical for Web commerce which can involve communicating with thousands of customers.

   (c) Another limitation with secret-key encryption is its inability to support non-repudiation. As both parties share the same key it is possible for one party to create a message with the shared secret key and falsely claim the other party had sent it.

   (d) Secret-key encryption on its own is not suitable for Web commerce - instead a system known as public-key encryption is used

2. **Public-key encryption :** Public-key encryption, or asymmetric encryption involves the use of two keys, one that can be used to encrypt messages (the public key) and one that can be used to either encrypt them or decrypt them (the private key).

These key pairs can be used in two different ways, to provide privacy or authentication.

Privacy is ensured by encoding a message with the public key as it can only be decoded by the holder of the private key.

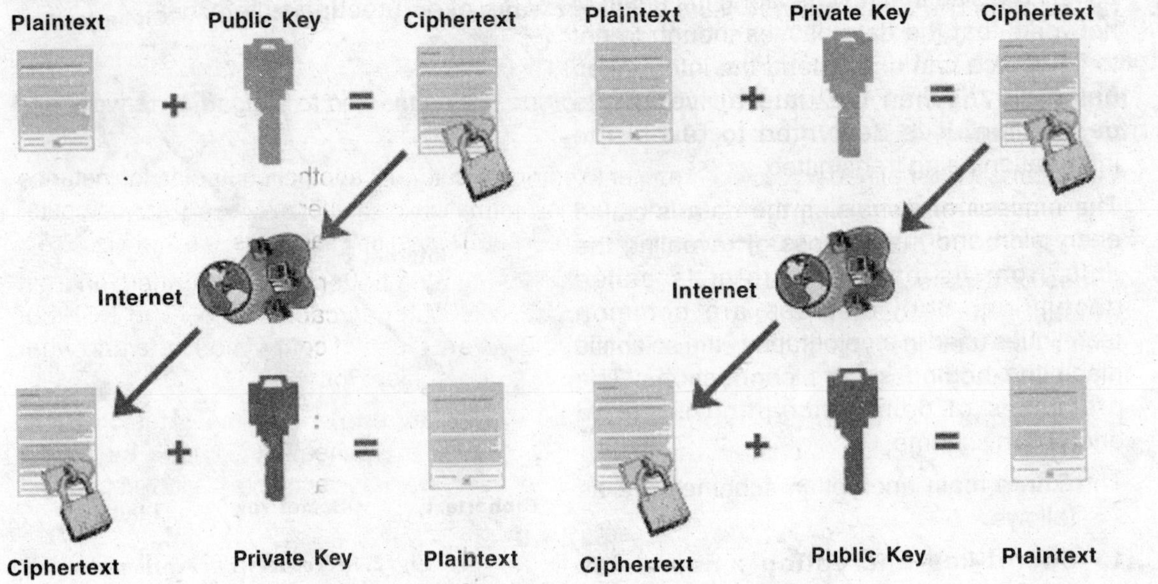

Fig. 3. Public Key Encryption.

Authentication is achieved by encoding a message with the private key. Once the recipient has successfully decrypted it with the public key they can be assured it was sent by the holder of the private key.

As the public key can be made widely available (for example from a server or third party), public-key cryptography does not suffer from the same key distribution and management problems as the secret-key system.

One disadvantage of the public-key system is that it is relatively slow, so when it is being used only for authentication it is not desirable to encrypt the whole message particularly if it is a long one. To get round this a digital signature is used.

2. **Digital signatures :** An increasingly important use for asymmetric encryption is digital signing. A digital signature is the reverse of public key encryption. Just like an ordinary signature it is used to prove the identity of the sender of a message. This can happen in several ways. The simplest is to send a random message as both plaintext and ciphertext. The recipient deciphers the ciphertext version using the published public key and if the two versions match it proves the sender was in possession of the private key.

   One drawback of this form of signature is that it only verifies itself, not any message to which it is attached.

   The RSA algorithm has become the standard for digital signature applications and the company founded by the co-inventors of RSA, RSA Data Security Inc, has developed (and published) a number of improvements on this basic scheme. Digital signatures on the Web are based on RSA, as is the popular PGP – Pretty Good Privacy – package.

Q. 4 (a). Explain the term Internet. Give different ways of connecting to Internet.

Answer 4(a)

Internet : ***The Internet*** *refers to millions of computers connected to a gigantic network and communicating via* ***TCP/IP protocols****.*

A protocol is a pre-defined way for a computer to communicate with another computer, for instance when requesting a service, or when forwarding some information to another machine. Each computer at any given time has a unique address on the Internet. This is its IP address.

Modes of Connecting to Internet : If we are looking for a better Internet connection then we must think about high-speed Internet connections, such as cable, ISDN, and DSL but need a little clarification as to what each of these different types of connections are and what they do. Read on, and I'll try to shed a little light on the subject for you.

1. **Telephone Line Internet Connectivity (Dial-up connection) :** The most basic type of Internet connection is called a dial-up connection. This connection is made through a modem (the communication mechanism in computers) that uses a telephone line to connect to the Internet.
2. **ISDN - Dial-Up Connection :** The second type of dial-up connection is through an ISDN (Integrated Services Digital Network). This connection is a high-speed dial-up and requires a special type of dedicated *digital* telephone line. This type of connection improves speed because the signals received are already in a digital format, but it is costly.
3. **Broadband Connections :** Broadband is a high-speed Internet connection that makes surfing the Web more enjoyable. It also easily accommodates the video, audio, or complex graphics that are becoming commonplace on the Internet. Broadband is a type of data transmission in which a single medium or wire can carry several channels or communication paths at once.
4. **Internet Through Cable :** Internet over cable is a technology and service that uses the existing cable TV networks (made up of coaxial cable lines that bring television signals to TV) to bring Internet data to PC or TV at a very high speed. The television and the Internet transmission take place simultaneously on the same cable but at different frequencies. This allows the user to view TV and access Internet at the same time.
5. **Internet Connectivity through Leased Line :** Leased lines are dedicated circuits provided by Basic Service Providers (BSPs), which provide permanent connectivity to the Internet. Leased lines provide the last mile access from the user premises to the ISP. They provide permanent connection as compared to the temporary connectivity through dialup access. The quality of the connection is far superior to what is normally available through dialup, thanks to digital signaling, less noise, fewer exchanges etc.

Q. 4. (b) Explain :

(i) Physical Component of Network

(ii) Internet Congestion

Answer 4(b) :

Physical Component of Network :

1. **Workstations :** All of the user computers connected to a network are called workstations. A typical workstation is a computer that is configured with a network interface card, networking software, and the appropriate cables.

2. **Network Interface Cards :** The Network Interface Card (NIC) provides the physical connection between the network and the computer workstation.
3. **Ethernet Cards :** Ethernet cards are usually purchased separately from a computer, although many computers (such as the Macintosh) now include an option for a pre-installed Ethernet card.

   Ethernet cards contain connections for either coaxial or twisted pair cables (or both) (See Fig. 1). If it is designed for coaxial cable, the connection will be BNC. If it is designed for twisted pair, it will have a RJ-45 connection.
4. **Token Ring Cards :** Token Ring network cards look similar to Ethernet cards. One visible difference is the type of connector on the back end of the card. Token Ring cards generally have a nine pin DIN type connector to attach the card to the network cable.
5. **Switch :** A network switch is a small hardware device that joins multiple computers together within one Local Area Network (LAN). Technically, network switches operate at layer two (Data Link Layer) of the OSI model.
6. **Repeaters :** Network repeaters regenerate incoming electrical, wireless or optical signals. With physical media like Ethernet or Wi-Fi, data transmissions can only span a limited distance before the quality of the signal degrades. Repeaters attempt to preserve signal integrity and extend the distance over which data can safely travel.
7. **Bridges :** A bridge is a device that allows you to segment a large network into two smaller, more efficient networks. If you are adding to an older wiring scheme and want the new network to be up-to-date, a bridge can connect the two.
8. **Routers :** A router translates information from one network to another; it is similar to a super intelligent bridge. Routers select the best path to route a message, based on the destination address and origin. The router can direct traffic to prevent head-on collisions, and is smart enough to know when to direct traffic along back roads and shortcuts.

Answer (ii):

Internet Congestion : Network congestion is the situation in which an increase in data transmissions results in a proportionately smaller increase, or even a reduction, in throughput.

Throughput is the amount of data that passes through the network per unit of time, such as the number of *packets* per second. Packets are the fundamental unit of data transmission on the Internet and all other TCP/IP (Transmission Control Protocol/Internet protocol) networks, including most LANs (Local Area Networks).

Congestion results from applications sending more data than the network devices (e.g., routers and switches) can accommodate, thus causing the *buffers* on such devices to fill up and possibly overflow. A buffer is a portion of a device's memory that is set aside as a temporary holding place for data that is being sent to or received from another device. This can result in delayed or lost packets, thus causing applications to retransmit the data, thereby adding more traffic and further increasing the congestion.

Q. 5. (a) What is FTP? Explain the working of FTP.

Answer :

FTP is a protocol to communicate over the network specifically designed for moving files between computers on the network. Primary use to most of us is transferring files to/from our desktop/laptop personal computers and the unix workstations, but it can also be used to communicate with machines at companies like Netscape to download software, or any PCs or other machines running FTP servers.

Using FTP to Download (Transfer from Another Machine to Yours)

1. **Connect to the Server (and Login) :** Servers with restricted access require a username and a password. Other servers (such as ftp.netscape.com) support "anonymous login" where the standard is to use the username "anonymous" and our full e-mail address as your password.
2. **Find the File(s)/Folder(s) We Want to Download :** Navigating the folder hierarchy is just like it is on PCs, etc. Double-click a folder to open it. (Command-line FTP programs are more like navigating in dos/unix; use the "cd" command).
3. **Choose the Appropriate Download Mode :** There are two transfer modes in FTP. The first is called either "ascii" or "text", and the other is called "binary" or "image". The reason for the two modes is that different machines represent End-Of-Line characters in different ways.
4. Begin the transfer and get the file (make sure we know where the file is being put on our local machine).

Using FTP to Upload (Transfer from Your Machine to Another)

1. **Connect to the Server (and Login) :** Servers with restricted access require a username and a password.
2. If we are using Fetch, choose to "Put" files/folders, and we will be prompted to specify the files.
3. Find the file(s)/folder(s) we want to upload on your computer. Navigating the folder hierarchy is just like it is on PCs, etc. double-click a folder to open it.
4. **Choose the Appropriate Upload Mode!** See the notes on download for details. Note that Fetch chooses different names for the file types when uploading. "Raw Data" is the same as binary/image. "Text" is text.
5. Begin the transfer.

Q. 5. (b). What is web browser? Explain different parts of a web browser.

Answer :

Pls. refer answer of question 3(a) in the solution of May-2007 paper

Q. 6. (a) Explain :

(i) Cookies

(ii) Collaborative Computing

Answer 6. (a) (i)

Cookie : A cookie is a computer text file sent to a visitor's Web browser (the software used to access the Internet such as Internet Explorer and Netscape) by a Web server (the computer that hosts the Website) in order to remember certain pieces of information. This can be a convenience for both Website visitors and operators because it can be used to reduce the amount of time to input and process the same information each time a Website is used. Information stored within a cookie can be read only by the Web server that originally sent the cookie, not by other Web servers.

Typically, a cookie comprises:

- a name for the cookie (chosen by the Website you are visiting)
- a value (unique number for the cookie) (determined by and stored by the Website for future recognition and action)
- an expiration date
- a valid path (details about the Web page(s) that the visitor was on when the cookie was sent)
- a valid domain (the name of the Website that created and can retrieve the cookie)
- a secure connection requirement (if the cookie is marked "secure", it will only be transmitted if the visitor is connected to a secure Website

There are two types of cookies :

1. **Session Cookies** : These cookies reside on the Web browser and have no expiry date. They expire as soon as the visitor closes the Web browser. Session cookies remember information only for as long as the visitor operates the Web browser in a single "session" (or "sitting"). Session cookies can be used by Website operators to determine information such as what parts of a Website are popular, how long people stay on certain sections of a Website and what browsers people are using.

2. **Persistent Cookies :** These cookies have an expiry date, are stored on a visitor's hard drive and are read by the visitor's browser each time the visitor visits the Website that sent the cookie. It is possible for the Website that created the cookie to extend the expiry date without notice to the visitor. They will remain there until the set date has expired or until the visitor has deleted the file.

Answer 6.(a)(ii)

Collaborative Computing : *Collaborative computing* is a term describing a variety of activities where people interact with one another using desktops, laptops, palmtops, and sophisticated digital cellular phones. As computers are best at handling data and representing information, person-to-person communication is enriched by an ability to share, modify, or collaboratively create data and information.

We believe collaborative computing is the future of the Internet. The Internet will evolve from its current role as a channel for information dissemination to a person-to-person communication medium. This is precisely where CollabWorx sees its role: we build, deploy, and support tools for personal interaction over the Internet.

It is the term that, we believe, will replace the traditional notion of "groupware." Groupware as we know it today focuses on the asynchronous aspect of human collaboration. E-mail, discussion lists, list servers, and similar software products help people to electronically exchange *information*. In other words, these products electronically extend a letter, a fax, and a filing cabinet.

Q. 6. (b). What is search engine? Explain the various components of Search Engine.

Answer :

Search Engine : Search engine is a software program that searches for sites based on the words that you designate as search terms. Search engines look through their own databases of information in order to find what it is that you are looking for. The Examples are Google, Yahoo, AltaVista, and Lycos.

There are four main components to a search engine which will affect the results of your search.

(1) The Webcrawler

This is the part of the search engine which combs through the pages on the internet and gathers the information for the search engine. Variable features which can affect your search results include :

- **Included Pages :** Most search engines will find information by beginning at one page and then following all of the links on that page. It will then follow all of the links on these new pages, and so on. Therefore, if a page is not linked to from another page, it may never be found by a search engine. Authors can include unlinked pages in a search engine by submitting them to each specific search engine.
- **Excluded Pages :** Some web administrators may choose to exclude their pages from search engines because they are internal pages or intranets. Many web pages are also excluded because their content is dynamically generated from a database and a search engine cannot find it.
- **Documents Types :** Different search engines will search different document types. All will search HTML documents, but some will also search PDF, PowerPoint, Word, Excel and more.
- **Frequency of Crawling :** An important part of a web crawler is how frequently it retrieves information from pages. Some sites it will visit more often than others.

(2) The Database

The search engine's database is what you are actually searching. All of the information that a web crawler retrieves is stored in a database. Every time you use a search engine, it is this database you are searching, not the live internet. Variable features which can affect your search results include :

- **Size of the Database :** Some search engines will have extremely large databases (~5 billion pages) for you to search, while others will have comparatively small ones (~150 million pages). A small database is not necessarily worse, however, since it may offer more focused and higher quality results then a larger database.
- **Freshness of the Database :** The freshness of the database is a direct result of how frequently the web crawler retrieves new information. If the information in the database is fairly old, then your search results will suffer.

(3) The Database

Each search engine interprets the terms you enter into the search box in different ways. Variable features which can affect your search results include :

- **Operators :** Most search engines allow you to use operators such as AND, OR and NOT in order to create complex search statements. The terms may need to be entered in upper case. Most also use the plus (+) sign to signal that a term must be included in the search results and a minus (–) sign to signal that a term must be excluded from the results.
- **Phrase Searching :** Search engines will generally search for words as phrases when marks are placed around the phrase. Some search engines may use a drop down menu which offers the option of searching the terms as a phrase.
- **Truncation :** Some search engines will automatically truncate the terms you enter. This means that the search engine will not only search for the term exactly as you spelled

it, but will also search on that term with alternate endings and as a plural. Some search engines will only search for variable endings on certain common words.

(4) The Ranking Algorithm

How a search engine ranks the results of your search is possibly the most important component of a search engine. Most searches will retrieve thousands of results. Since you probably will only look through the first 1-2 pages of results, you need the most relevant results to appear first. Variable features which can affect your search results include :

- **Location and Frequency :** All search engines look at the location and frequency of words in a page. If a term appears near the top of a web page, such as in the title or in the first few paragraphs of text, it is assumed that the page is more relevant than if the term is used at the bottom of the page. Pages where the words appear more frequently in relation to the other words on the page also qualifies the page as being more relevant than other web pages.
- **Link Analysis :** This feature analyzes how pages link to each other and then uses this information to determine the "importance" of each page. If a page is linked to from a large number of other pages, then it is ranked more highly.
- **Clickthrough Measurement :** Some search engines also use clickthrough analysis. This means that a search engine might watch what results someone selects from a particular search, then eventually drop high-ranking pages that aren't attracting clicks, while promoting lower-ranking pages that do pull in visitors.

Q. 7. (a). What is cascading style sheets ? How they are implemented ?

Answer 7. (a) :

Style sheets are a very powerful tool for the Website developer. They give you the chance to be completely consistent with the look and feel of your pages, while giving you much more control over the layout and design than straight HTML ever did.

Invented in 1997, Cascading Style Sheets (CSS) are just now starting to be widely used among browsers and Web developers are learning to be more comfortable with them. Those of you who use HomeSite 4.0 know that they are eventually going to take the place of tags such as <font>, which have been deprecated in HTML 4.0.

There are three parts to CSS :

- The styles,
- Their placement, and
- The fact that they can cascade.

Implementation CSS Styles : Cascading Style Sheets (CSS) are a collection of formatting rules that control the appearance of content in a web page. They are very useful for maintaining a website since its appearance (controlled by properties of HTML tags) can be managed from just one file. CSS Styles also enhance your site's look, accessibility and reduces file size. Another main advantage is reusability - instead of defining the properties of fonts, backgrounds, borders, bullets, uniform tags, etc. each time you use them you can just assign the corresponding css style in the class property. You can store CSS styles directly in each document or, for more control and flexibility, in an external style sheet.

Q. 7. (b) Explain various features of XML.

Answer (b) :

Pls. refer answer of Question 3(b) in the solution of May-2007 paper

Q. 8. Explain :

(i) **Digital Signature**

(ii) **DNS**

(iii) **IIS**

(iv) **Server Side Programming**

Answer :

Digital Signature : A **digital signature** is basically a way to ensure that an electronic document (e-mail, spreadsheet, text file, etc.) is **authentic**. Authentic means that you know who created the document and you know that it has not been altered in any way since that person created it.

Digital signatures rely on certain types of **encryption** to ensure authentication. Encryption is the process of taking all the data that one computer is sending to another and encoding it into a form that only the other computer will be able to decode. Authentication is the process of verifying that information is coming from a trusted source. These two processes work hand in hand for digital signatures.

DNS : The Domain Name System (DNS) is the way that Internet domain names are located and translated into Internet Protocol addresses. A domain name is a meaningful and easy-to-remember "handle" for an Internet address.

Because maintaining a central list of domain name/IP address correspondences would be impractical, the lists of domain names and IP addresses are distributed throughout the Internet in a hierarchy of authority. There is probably a DNS server within close geographic proximity to your access provider that maps the domain names in your Internet requests or forwards them to other servers in the Internet.

Internet Information Server (IIS) : IIS is a set of Internet-based services for servers created by Microsoft for use with Microsoft Windows.IIS comes with Windows 2000, XP, and Vista. It is also available for Windows NT. IIS is easy to install and ideal for developing and testing web applications.

IIS includes Active Server Pages (ASP), a server-side scripting standard that can be used to create dynamic and interactive web applications. It is the world's second most popular web server in terms of overall websites behind the industry leader Apache HTTP Server.

How to Install IIS on Windows XP and Windows 2000

Follow these steps to install IIS on Windows XP, and Windows 2000:

1. On the Start menu, click Settings and select Control Panel
2. Double-click Add or Remove Programs
3. Click Add/Remove Windows Components
4. Click Internet Information Services (IIS)
5. Click Details
6. Select the check box for World Wide Web Service, and click OK
7. In Windows Component selection, click Next to install IIS

Server Side Programming :

Server side programming is used to centrally manage all programs relating to a particular application (e.g. Banking, Insurance, e-shopping, etc). Clients with bare minimum requirement (e.g. Pentium II, Windows XP Professional, MS Internet Explorer and an internet connection) can experience the power and performance of a Server (e.g. IBM Mainframe, Unix Server, etc.). from a remote location without having to compromise on security or speed. More importantly, server programs are not only portable but also possess the capability to generate dynamic responses based on user's request.

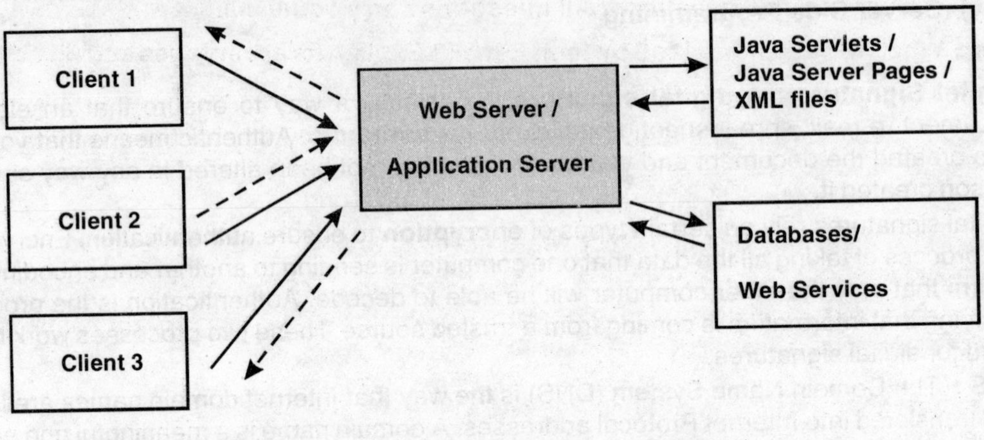

Fig. 4. Server Side Programming (Architecture)

Advantages of Server Side Programs

The list below highlights some of the important advantages of Server Side programs.

(i) All programs reside in one machine called the Server. Any number of remote machines (called clients) can access the server programs.

(ii) New functionalities to existing programs can be added at the server side which the clients' can advantage without having to change anything from their side.

(iii) Migrating to newer versions, architectures, design patterns, adding patches, switching to new databases can be done at the server side without having to bother about clients' hardware or software capabilities.

(iv) Issues relating to enterprise applications like resource management, concurrency, session management, security and performance are managed by service side applications.

(v) They are portable and possess the capability to generate dynamic and user-based content (e.g. displaying transaction information of credit card or debit card depending on user's choice).

2. Types of Server Side Programs

(i) Active Server Pages (ASP)

(ii) Java Servlets (JS)

(iii) Java Server Pages (JSPs)

(iv) Enterprise Java Beans (EJBs)

(v) PHP

FOURTH SEMESTER EXAMINATION, May 2008

INTERNET FUNDAMENTALS (CSE-208-E)

Note : Attempt any five questions. All questions carry equal marks.

Q. 1. (a) What do you understand by term E-mail? Explain its advantages and disadvantages.

Answer 1 (a)

E-Mail : You can send e-mail from your computer at any time of the day to any address around the world and your electronic letter will arrive at its destination seconds after you send it, even if the receiver lives on the other side of the world.

"E-mail is as simple and straight forward as letter writing. E-mail is an electronic version of sending a letter".

"E-mail, e-mail or electronic mail is the transmission of messages (e-mails or e-mail messages) over electronic networks like the internet".

Advantages of E-mail : Electronic mail, or **e-mail**, is the most frequently used service on the Internet for many reasons:

1. Send a message **anytime**, **anywhere** and the recipient can read it at his or her convenience.
2. Send the same message to **multiple recipients**.
3. **Easy to use**, E-mail frees us from the tedious task of managing data of daily use. It helps us to manage our contacts, send mails quickly, maintain our mail history, store the required information, etc.
4. **Forward information** to people without retyping it.
5. **E-mail is fast**, usually taking no more than a few minutes to be received.
6. **Attach digital files** to your messages, including electronic documents, video clips, music and photos.
7. **Send messages around the world** as easily as to someone down the block.
8. Sending and receiving e-mails is typically free or having **very low cost.**

Disadvantages of E-Mail : Like other good thing e-mail also has some disadvantages as follows:

1. E-mail reaches the recipient most of the time, but **delivery is not guaranteed**. Sometimes spam filtering software may block a message or direct it to the recipient's **Junk Mail folder**.
2. There is **no guarantee** of e-mail privacy. A message can be intercepted and read as it makes its way over the Internet to its final destination. Consider e-mail as you would a post card.
3. Unencrypted mails are unsafe. So the safest way to send sensitive or confidential information via e-mail is by using **encryption software.**
4. Unsolicited e-mails that are sent in huge numbers and clog mail servers as well as e-mail programs.
5. With hundreds of these junk mails in our inbox, the occasional good e-mail can be hard to spot.

6. The recipient needs to scan the mails, as **viruses** are transmitted through them and have the potential to harm computer systems.
7. Checking and deleting these unwanted mails can unnecessarily consume a lot of time, and it has become necessary to block or filter the unwanted e-mails by means of ***spam filters.*** Spamming includes, sending hoax e-mails. *E-mail spoofing* is another common practice, used for spamming. Spoofing involves deceiving the recipient by altering the e-mail headers or the addresses from which the mail is sent.

Q. 1. (b) Explain:

(i) Mailers

(ii) Mail Servers

(iii) Mail boxes

Answer :

Mailers : A mailer is a company that sends direct mail. A mailer often employs a list broker to assist with buying mailing lists and a service bureau to assist with the list processing necessary to conduct their direct marketing campaigns. The term mailer is sometimes used to refer to a marketer sending e-mail or other direct communications.

Types of Mailers : There are many different types of mailers and each type of mailer is available in a variety styles. Different types of mailers are used for different purposes, are manufactured by different types of printing processes and are processed in different ways. Shown below are two types of mailers:

1. **Promotional Mailers :** Promotional mailers are used for advertising and promotional purposes. They generally have more of a creative design to attract the customer's attention to the product or services being advertised. The purpose of the mailer may be to advertise special offers or discounts on products or services, inform customers about new products and services, distribute company information or advertise industry conferences and training seminars. The type of mailer used can vary from a single postcard to a folded multiple page mailer.
2. **Self Mailers :** Self mailers are generally used for business applications such as statements, checks, order forms, billing notices, credit memos, past due notices, proxy notices, licensing, correspondence and many others. The mailer may be a one way mailer that contains information sent to the recipient without a response request or it may be a business reply piece only, such as a credit application or an order form from a catalog. The mailer could also be a two way mailer that sends information to the recipient and also contains a response mail piece requesting a response from the recipient. Self mailers save time and handling costs by eliminating the need for a mailing envelope, which improves the efficiency of the mailing process. The self mailer can contain a reply envelope which promotes a quicker response time from the customer.

Mail Servers : A MailServer is a Web server set up to handle e-mail by having two programs installed: a *sendmail* program and a *POP* program. POP (Post Office Protocol) collects mail sent to you. A mailserver can be installed on the main Web server, or on a separate server.

A mail server (also known as a *mail transfer agent* or MTA, a *mail transport agent*, a *mail router* or an *Internet mailer*) is an application that receives incoming e-mail from local users (people within the same domain) and remote senders and forwards outgoing e-mail for delivery.

A computer dedicated to running such applications is also called a mail server. Microsoft Exchange, qmail, Exim and sendmail are among the more common mail server programs.

Mail boxes : In electronic mail, the storage space that has been set aside to store an individual's electronic mail messages. In e-mail systems, each user has a private mailbox. When the user receives e-mail, the mail system automatically puts it in the mailbox.

Q. 2. (a) Explain the term Internet. Give its various advantages.

(b) Explain:

(i) Physical Components of network

(ii) Internet Congestion

Answer :

Kindly refer the solution of question number 4 in December -2008 paper

Q. 3. Write an HTML Program using 10 different HTML tags.

Answer: Kindly refer the solution of question number 2-a in December -2008 paper

Q. 4. (a) What is Search Engine? Explain various components of search engine.

Answer :

Kindly refer the solution of question number 6-b in December -2008 paper

Q. 4. (b) What is FTP ? Explain the working of FTP.

Answer :

Kindly refer the solution of question number 5-a in December -2008 paper

Q. 5. (a) What is cascading style sheets? Explain Inline, Internal and External styles.

(b) Explain XML in detail.

Answer :

Kindly refer the solution of question number 7 in December -2008 paper.

Q. 6. (a) What do you understand by software complexity? What are its contributing factors ?

Answer :

Software Complexity : Complexity is a measure of the resources expended by a system while interacting with a piece of software to perform a given task. If the interacting system is a computer, then complexity is defined by the execution time and storage required to perform the computation. If the interacting system is a programmer, then complexity is defined by the difficulty of performing tasks such as coding, debugging, testing, or modifying the software. The term software complexity is often applied to the interaction between a problem and a programmer working on some programming task.

Q. 6. (b) Differentiate between Public key cryptography and Private key cryptography.

Answer :

Private Key cryptography : Private key encryption is the standard form. Both parties share an encryption key, and the encryption key is also the one used to decrypt the message. The difficulty is sharing the key before you start encrypting the message - how do you safely transmit it?

Many private key encryption methods use public key encryption to transmit the private key for each data transfer session.

Public Key cryptography : Public key encryption uses two keys - one to encrypt, and one to decrypt. The sender asks the receiver for the encryption key, encrypts the message, and sends the encrypted message to the receiver. Only the receiver can then decrypt the message - even the sender cannot read the encrypted message.

Q. 7. (a). What is web browser? Explain different parts of web browser.

Answer : Kindly refer the solution of question number 3-a in May -2007 paper

Q. 7. (b) Differentiate between :

(i) IP v4 and IP v6

(ii) Plug-ins and Helper

Answer :

(i) Kindly see the section 3.10.3 of Chapter 3 of the book.

(ii) Kindly see the section 6.4.1 for Chapter 6 of the book.

Q. 8. Explain :

(i) Firewall **(ii) Internet Service Provider**

(iii) Web Servers **(iv) Server Side Programming**

Answer :

Firewall : Kindly refer the solution of question number 7-b in May -2007 paper

Internet Service Provider : Kindly refer the solution of question number 8 in December-2008 paper

Web Servers : Web server is the program resides on server computer to server the request of the client software.

Web servers are a combination of software and hardware that store documents and other content. The server content is accessed using Web browsers. Web servers are available for a wide range of platforms including IBM PCs, Apple Macintosh computers, UNIX platforms and a large number of proprietary hosts.

Some the example of the web servers are as follows :

1. Tomcat
2. Weblogic
3. Apache
4. IIS

Server Side Programming : Kindly refer the solution of question number 8 in December-2008 paper.

❑❑❑

INDEX

INDEX